DESIGNER UNIVERSE

THE world is charged with the grandeur of God.
    It will flame out, like shining from shook foil;
    It gathers to a greatness, like the ooze of oil
Crushed. Why do men then now not reck his rod?
Generations have trod, have trod, have trod;
    And all is seared with trade; bleared, smeared with toil;
    And wears man's smudge and shares man's smell: the soil
Is bare now, nor can foot feel, being shod.

And, for all this, nature is never spent;
    There lives the dearest freshness deep down things;
And though the last lights off the black West went
    Oh, morning, at the brown brink eastward, springs—
Because the Holy Ghost over the bent
    World broods with warm breast and with ah! bright wings.

                  'God's Grandeur', G. M. Hopkins

To Vicky, Elizabeth, George and Charles

DESIGNER UNIVERSE

**Is Christianity compatible with
modern Science?**

John Wright

MONARCH
Crowborough

First published 1994

Unless otherwise indicated, biblical quotations are from
the Revised Standard Version

ISBN 1 85424 260 1

British Library Cataloguing in Publication Data
A catalogue record for this book is available
from the British Library.

Designed and Produced in England for
MONARCH PUBLICATIONS
Broadway House, The Broadway, Crowborough,
East Sussex TN6 1HQ by
Nuprint Ltd, Station Road, Harpenden, Herts AL5 4SE.

Contents

Acknowledgement

The quotation from *The Selfish Gene* by Richard Dawkins (OUP, 1989) is used by permission of Oxford University Press.

Preface

FOR THE ORIGINS of this book one must go back a dozen years or so. At that time, my then Rector, the Rev John Skinner, asked me to give a short talk on science and Christianity to a group of young people and I did not find the task an easy one. This was partly because the subject is complex, but I suspect also that the maxim 'If you can't explain something clearly it is because you yourself do not really understand it' came in as well. Feeling rather inadequate I resolved to treat the issues more thoroughly when time permitted. This book is the result.

But the book is also more timely today than it would have been twelve years ago with modern scientific developments in fields such as cosmology, chaos theory and artificial intelligence receiving considerable public interest. If Stephen Hawking can write a best seller on cosmology in which he comments on the role of God in creation, should not we Christians, who have an appreciation of the science involved, have something to say? Although science may no longer be as popular as it once was, nor scientists appear as omniscient as they once did, we all recognise the significance of science and technology in our lives. Moreover, it is exciting. I still find it as fascinating as I did when I was studying Physics at Cambridge some forty-five years ago. Although there are still areas of uncertainty, for reasons explained in the book, many of the apparent conflicts between science and Christianity have now been resolved. Many of the great scientists have expressed a sense of awe at the beauty and harmony of the universe revealed to them through their science. Christians can take the further step of attributing the wonder of creation to the great God they love and serve.

We live in an age when it is customary to explain science in a popular way. This book is no exception. It covers a very wide range of science at a fairly basic level and it is hoped that readers will find this valuable. But the real objective is to show how all this fits in with Christianity. In keeping with the spirit of the book I have tried to set the discussion at as simple a level as possible while addressing the real issues. It is intended for the general reader, not professional theologians or philosophers.

In undertaking the work, I have had a great deal of encouragement and support from many people. First of all I am grateful to the clergy and members of my own church, St Saviour's, Guildford. Some time ago I did some work on a Christian view of Chernobyl with a group at Luton Industrial College. When the Director, the Rev Harold Clarke, heard that I was intending to write this book, he invited me to centre my activities there. I have had tremendous encouragement and constructive criticism from Mr Clarke as well as other members of staff. In particular I am grateful to Miss Madeleine Major who typed (and retyped and retyped!) half the manuscript.

I am also indebted to Marguerite, my wife, for both her encouragement and understanding and for typing the other half of the manuscript and producing the drawings.

Several people have taken the trouble to read and provide constructive comments on the draft. In particular I am indebted to Dr Oliver Barclay, the Rev Dr Richard Burridge, Mr Douglas Gutherie, the Rev Dr David Hardy, Professor Tom Torrance and my daughter, Mrs Jane Bramhall, JP.

The Rev Dr Donald English, who in his role as Secretary of the Home Mission Division of the Methodist Church has Luton Industrial College as one of his many responsibilities, was kind enough to find the time to write a Foreword.

Foreword

THERE IS AN INCREASING WILLINGNESS, on the part of scientists who are Christians, to address issues raised in the area of science and faith. The mood is much less confrontational. There is an openness about the parallel nature of much of the work done by scientists on the one hand and theologians on the other. There is also both courage and perception in the identification of areas of continuing disagreement.

This book is a splendid contribution to the process. John Wright shows an impressive grasp of a wide range of scientific disciplines, and he identifies most helpfully the major steps in the development of modern scientific knowledge. What is more, he does it clearly and simply, so that non-specialists can read and understand. The combination of history of science and scientific explanation is most instructive and helpful.

Dr Wright also works on the basis of a firm personal Christian faith. He is committed to the biblical and theological content and implications of Christian believing. There is also here a very sensitive distinction between that which may be firmly held and clearly understood, and that which is simply part of the mystery of faith.

In both areas of description and evaluation, therefore, this book provides clarification and perspective which enable the conversation to continue. Above all, from the scientific point of view it is clear that Christian faith is not excluded, and that from the Christian point of view there is nothing to fear from well-founded scientific conclusions.

Many people will find this book extremely helpful. John Wright is to be congratulated on such a perceptive and timely piece of work, as is Luton Industrial College for its wisdom in encouraging its production and publication.

The Rev Dr Donald English
Chairman, World Methodist Council

1

The God of Science

'Religion without science is blind,
Science without religion is lame.'
Albert Einstein

About this book

W E LIVE IN AN AGE that is increasingly dominated by
science. From the dawn of a new, scientific way of
thinking some three hundred years ago, our knowledge of how the universe works has progressed at an ever-
increasing pace. The technological advances we have lived
through during the last fifty years–radar, nuclear energy,
space travel, television, computers, plastics, antibiotics, and
the increasing use of technology in medicine–show how
science has affected all of us in our daily lives. This view of
the progress and power of science has penetrated so deeply
into our culture that we have come to expect dramatic
advances and, moreover, tend to accept what we are told
about them without question when they do occur.

At the same time we live in an age when increasing
numbers of people are finding a Christian faith that meets
their deepest personal needs. They have responded in faith
to an invitation to entrust their lives to one they cannot see
but who claims to be the Way, the Truth and the Life. Very
often they find it hard to explain why they responded as they
did, but they find their lives changed as a result. In many
cases the response has been made without prior deep intel-

lectual thought. There is nothing wrong with this. Jesus looks for trust, not cleverness. But often, as time progresses, the new believer comes face to face with doubts—and some of these could be centered on the relationship between science and Christianity. Does one disprove the other? This book is written with such people in mind.

Others find they need to face up to such issues before they can make a full Christian commitment. This book is intended to help those in this situation.

Finally there are those who already have a strong Christian faith and no serious doubts of their own. But they wish to commend Jesus to others and, in discussion, the question of science has arisen. This book should help them understand the issues so that they can explain them more clearly.

In writing a book such as this, one has to consider how deep and technical it should be. There are already a number of excellent modern books on the subject, but not all of them are easy to read. For those who wish to study the subject in depth, the four books by Polkinghorne are excellent. In contrast, this book covers the ground as simply as possible but without being over simplistic. It is written for the non-scientist but it should also appeal to the scientist. Those who wish to consider some of the scientific background in more depth will find it covered in the Appendices. While the science has been written at a level comparable with that of popular TV programmes such as *Horizon* and *Equinox*, a deliberate decision has been taken to present the theological and philosophical issues at a somewhat simpler level so they will be readily understood by the average intelligent layman.

In the past many Christians have seen science as something of a threat. This is surely wrong. Science is basically a systematic way of exploring God's creation. The results are often surprising and beautiful. They reveal a universe that is both complex and finely tuned. If the laws of science had been only very slightly different from what they are, life in the advanced form we know it, could not exist. We live in a 'Designer Universe'. Instead of adopting a negative, defensive attitude to science, we should be praising God for the wonders of his creation that are revealed to us through it. This book is intended to help the reader take this positive approach with intellectual honesty. As a first step towards this, we must remind ourselves of some of the relevant things the Christian believes about God and man, and the interaction of these beliefs with science.

The God of science

Before proceeding to look at the interaction between science and Christianity, it is helpful to remind ourselves of some of the relevant key features of what we believe. These fundamentals are 'not negotiable'. They are firmly believed by the vast majority of Christians throughout the world, in earlier generations as well as our own. They were summarised by the early church in its creeds, an important one, the Nicene Creed, being formally adopted in the fourth century. It starts: 'We believe in one God, the Father, the Almighty, maker of heaven and earth, of all that is seen and unseen.'

1) One God
Before the universe existed there was God–Father, Son and Holy Spirit. God was there before the 'big bang'. He was there before he created the 'laws of nature' on which all science depends. As such he was before time and space and stands outside time and space as well as being involved in it. We shall look in more detail at modern views of the universe and its creation in Chapter 3. But for now we just note that Einstein's General Theory of Relativity links matter, space and time in one package. We humans are therefore necessarily limited in our scientific knowledge to events that have happened within the lifetime of the universe in which we live–a period believed to be about 15 billion years. God has no such limitation.

Although God is almighty, he does permit himself to be constrained to some extent. He does this, for example, when he continues to give man free will–a topic we will return to in Chapter 6. But for now we will just consider how he has chosen that there should be rules governing the physical behaviour of the universe and that they should not normally be violated. We often call them the Laws of Nature although they should, more accurately, be described as part of God's Laws. When he is embarking on his research, the scientist consciously or unconsciously assumes that there is a pattern in the universe and sets out to establish what it is. He would be very reluctant to come up with a theory suggesting that the laws vary from one part of the universe to another or that they change with time. In making this assumption, the scientist is completely in line with the Christian belief in one eternal unchanging God who controls the whole universe in every detail.

As an example, our modern scientific views on the origin of the universe are largely based on one set of observations of distant galaxies; namely, the fainter the galaxies are, the more 'red' they appear. An explanation could possibly have been drawn up along the lines that the faintest galaxies are furthest away and different laws of physics apply in these remote regions. As we have seen, a scientist would be very reluctant to accept such a theory. Then, in 1929, the American astronomer, Hubble, pointed out that because of the Doppler Effect (the phenomenon in acoustics that causes the pitch of the sound of an approaching car to fall as it passes us, or the equivalent optical effect that causes a fast moving light source to appear less blue and more red as it passes us) the phenomenon could simply be explained if the universe was expanding, with the distant galaxies moving away most rapidly.

This constancy of the 'laws of nature' is consistent with the Christian belief in one God. Of course it in no way 'proves' the existence of God. It is a mistake to think that you can prove that God exists in the same way that you can prove a mathematical theorem. But it is one factor to be taken into account when we claim that our faith is firmly based.

2) The Father, the Almighty

We have just seen how God chooses to act within laws, the 'Laws of Nature', which he has prescribed. There is a further way in which he chooses to limit himself–by giving men freewill. We are in a deep theological area here. In essence, God is almighty but he seeks to win the love of his human creation, not through force but by loving us. This is of course shown most clearly in the life and death of Jesus Christ

> who, though he was in the form of God, did not count equality with God a thing to be grasped, but emptied himself, taking the form of a servant, being born in the likeness of men. And being found in human form he humbled himself and became obedient unto death, even death on a cross (Phil 2:6-8).

Here we are in an area where science can say very little. One can ask questions, which are partly scientific, about how, if God constrains himself by his laws of nature, can we have free will and how can he act in the world. This is a key issue and is considered in detail in Chapters 2 and 6. But at this stage it is important to note that there are some issues on which science can say very little even though they are just as real as those that can be explored scientifically.

Let us consider a very simple illustration. There are many excellent works of art depicting the crucifixion of Jesus. A physicist or chemist could examine such a picture and determine the composition of the paint that had been used, the nature of the canvas, possibly an indication of the date that it had been painted. A medical scientist could say that a man in that situation would not live long. Both of these are true. But to suggest that this is the whole story is to miss the point. The Christian would see a much deeper meaning. 'That is a picture of my Lord. He loved me so much that he did that for me.' There is no contradiction–the scientific view and the theological view are both true but complementary. The scientist would, in his sphere, be able to provide a detailed analysis that the theologian would be foolish to question. Equally, the purpose of Christ's death is a question for faith rather than science. In the past there has been so much misunderstanding caused by scientists believing they could speak with authority on theological or philosophical issues, and vice versa.

There is, of course, the possibility of overlap between the theological and scientific views. Thus, if the scientist dated the picture and came to the conclusion that it had been painted more than 2000 years ago, that would give problems! The test is therefore whether the theological and scientific views of God's world are consistent with each other.

We cannot expect complete consistency. Our almighty God does not reveal himself completely to his creation. Although he has revealed his purposes in Jesus, these need to be interpreted in the culture of every generation. 'We see through a glass darkly' (1 Cor 13:12, AV).

Equally, science is not static. Our knowledge of the world we live in develops and is refined. In some cases our current views are quite different from those of a hundred years ago. Nevertheless, we should be looking for a high degree of consistency between our theological and scientific views of the universe.

Notice that in the last few paragraphs the word 'theological' has been used to describe a view which is contrasted with a scientific one. The word 'Christian' was quite deliberately not used. Christ is Lord of all–and this includes science as well as religion.

Returning to consider 'our Father, the Almighty', the fatherhood of God implies one who cares for each of his children. He is the one to whom we can pray. He is the one who has a plan for our lives, who guides us through the Holy Spirit. He is a holy God who expects his standards to be displayed in the life of his children. He is a loving and forgiving Father who rejoices when men turn to him. Our God is active in the world–one who, at the end, will see his purposes worked out. Our God reigns!

For the Christian all this is just as real as any scientific discovery. Indeed, in a way it is like one. The scientist looks at the world, makes observations and develops a theory that will be accepted by his colleagues if they feel deep down that it is right. The theory may have to be refined or even rejected if new evidence comes to light that it cannot account for. There is no ultimate proof of a scientific theory. It remains valid only so long as fresh evidence continues to support it. The acceptance of a scientific theory depends, at the end of the day, on whether the scientific community judges it to have an elegant simplicity that makes it feel right. Similarly, our Christian faith is fostered by a community, the church which is itself under the authority of Scripture. But at the end of the day, our individual Christian belief hinges on whether, in the light of our experience of life in its widest sense, it brings to us a meaning, purpose and cohesion which we judge deep down to be true and real. 'For all who are led by the Spirit of God are sons of God. For you did not receive the spirit of slavery to fall back into fear, but you have received the spirit of sonship. When we cry, "Abba! Father!" it is the Spirit himself bearing witness with our spirit that we are children of God' (Rom 8:14-16).

3) *Maker of heaven and earth, of all that is seen and unseen.*
This section of the creed talks about God the creator. We must distinguish between God's original creative act when he brought the universe into existence from nothing, and his subsequent creative acts. So often we concentrate our attention on God the creator of the 'big bang' at the start of time; or God the creator of the first life on earth some three or four billion years ago. These are, of course, key occasions. But the creed talks about God making all—both seen and unseen. We must never forget that God is actively involved in all of nature, all art, music and literature, and all creative breakthroughs in science. Of course I cannot prove this—it is part of my Christian faith just as much as when I believe, but cannot prove in a rigorous logical sense, that God is active in

my life. The hidden God is there, and countless millions of Christians through the ages can tell of their experience of him. Similarly in all of creation, the hidden God is there, actively working.

As scientific knowledge has increased, so the mechanisms of creation have become clearer. In the past Christians had considered that God's creation is so complex and so wonderful that God alone, unaided and working outside his normal laws must have engineered it. So when Darwin suggested that evolution and natural selection were important mechanisms in biological development, many Christians felt (and some still feel) that God was being excluded. Of course he was not! Can't evolution and natural selection be part of God's plan and can't God be active in it? We will return to this topic again and again.

At this point I cannot help being reminded of the story of the Christian who went for a walk on the sand at the seaside and found himself cut off by the tide. In his predicament he prayed and felt sure God would save him. After a while, when the sea was up to his knees, a helicopter arrived and let down a man to help him. 'It's all right,' said the Christian, 'God will save me.' So the helicopter went away. A little while later, by which time the sea was up to his chest, a fishing boat arrived and the crew asked if he needed help. 'No it's all right,' said the man, 'God will save me.' A little while later he drowned.

On arriving in heaven the man met Jesus and said, 'I was so sure, Lord, that you were going to save me.' Jesus replied, 'Didn't I send a helicopter and didn't I send a fishing boat?'

So often we get confused and tend to believe that God only acts through 'spiritual' means, not 'natural' ones. When we stop and think we realise this is nonsense – God is the maker of all that is seen and unseen.

So when we pray for healing for someone who is sick, we may be tempted to feel that God is more active if the healing is sudden and without the aid of doctors than if the person is healed through the normal medical route. Of course God is equally active whichever way he chooses to deal with the problem. So it is with God's creative action in all aspects of the world. If we find a scientific explanation for some phenomenon, it does not mean that God is not acting–all that it means is that he has permitted us to see something of himself in action through the medium of science.

This is a key issue and we shall return to it many times in this book. It is an issue that has been hotly debated over the last four hundred years. It formed the basis of the first major challenge to Christian thought by science. In the next chapter we shall examine the nature of this challenge and its resolution in the light of modern scientific knowledge.

2

Is the World Like a Machine?

*'There are more things in heaven and earth, Horatio,
Than are dreamt of in your philosophy.'*
                    Hamlet, William Shakespeare

An early confrontation between science
and the church

MAN HAS ALWAYS been fascinated by the stars. It is clear that the heavens were mapped from the earliest times for which we have records. There may have been several reasons for this. There was the very practical reason of navigation in the Mediterranean. Then the motion of the sun and the stars could be used to show where one was in the seasons of the year and enabled the production of calendars. But there was also an elegance and rhythm in the motion of the stars which could be described mathematically.

The development of mathematics probably started as early as 2000 BC in Babylon and Egypt but flourished in Greece under the leadership of Pythagoras (around 580 BC). Another famous Greek mathematician was Euclid, who set up a school of mathematics in Alexandria, on the Nile, around 300 BC. Euclid was the man who systematised geometry and his book *Elements of Geometry* was translated and copied more than any book other than the Bible, right into modern times. This development of mathematics enabled the motion of the sun and stars to be described in mathematical terms so that accurate forecasts could be made of their future positions. The high point of all this work came when another Greek, Claudius Ptolemy, working in Alexandria around AD 150, showed that the motion of the planets could be accurately described in terms of epicycles. He assumed that there is a stationary earth at the centre of the universe. The planets move around in circles, the centres of which are simultaneously moving around larger circles. This model was so accurate that it was not challenged fundamentally for 1400 years. It appealed to the Greeks because they believed that the perfect form of motion is a circle. It appealed to the medieval church because it considered God's highest creation, man, should naturally be on a special body, the earth, which was at the centre of the universe.

The challenge, when it came, was part of the intellectual, political and religious upheaval that we know as the Renaissance and the Reformation. Three individuals stand out as playing a crucial role in the debate. The first is Nicolaus Copernicus, a distinguished Pole, who was born in 1473. He was involved in the administrative affairs of his time–advising his government on currency reform and the Pope on calendar reform. Copernicus was struck by the apparent complexity of the epicyclic motion of the planets. Surely nature should be simpler than that? For years he considered the issue and eventually, in 1543, just before his seventieth birthday, he published his mathematical description of the heavens *The Revolution of the Heavenly Orbs*. The key feature was that he envisaged the sun sitting at the centre of the universe with the earth and planets revolving around it. At a stroke he removed the complexities of the epicyclic model and introduced a much simpler one.

To us in the twentieth century, this idea does not seem particularly startling. But in the sixteenth century it was dynamite. The church had almost come to accept the system of Ptolemy as an article of faith. After 1400 years it had become ingrained in medieval culture. It is not of course a real challenge to Christianity, but merely to the culture of the times. In our generation we must be careful to distinguish between what Christ is saying to us from the presuppositions that we have inherited from our cultural background. At the same time we must recognise that such distinctions are not easy to make and we must take care not to 'throw the baby away with the bath water' when we are questioning our traditions.

The second prominent individual is Johannes Kepler. He had available to him a series of new and highly accurate astronomical observations that had been made by Tycho Brahe, and he made a careful mathematical analysis of them. Kepler found that planets move in elliptical (not circular) orbits around the sun. He also established the relationship between the speed of the planet and its distance from the sun. (The line joining the planet and the sun sweeps out equal areas in equal times. He also found that the period of rotation of a planet in its orbit is proportional to the three halves power of the distance from the sun.) These were fascinating observations. But it was over a hundred years before Newton would explain these results.

The third outstanding figure is that of Galileo. Born in the same year as William Shakespeare (1564), Galileo Galilei rapidly built a reputation in Pisa as both an outstanding scientist and as a maker of scientific instruments. He is often regarded as the first scientist in the modern sense of one who builds apparatus and experiments to establish the way the world works. He is remembered for his experiments on gravity (those involving the timing of cylinders rolling down inclined planes were more significant than the well-known story of weights dropped from the leaning tower of Pisa) and for building the most powerful telescope of his time (magnification thirty). He discovered the satellites of Jupiter and produced the first maps of the moon's surface. His observations convinced him that the Copernican view of the universe was correct and the traditional Ptolemaic view was wrong. In an age when the authority of the Roman Catholic Church was having to be vigorously defended against heretics of all kinds, to openly discuss Copernicus' book was extremely dangerous. Galileo had become famous after the publication of his astronomical findings *Siderius Nuncius* (The Starry Messenger) in 1610. For such a famous man to publish a further book in 1632 discussing the orthodox and Copernican models, *The Dialogue on the Great World Systems* was too much for the Pope–even an intellectual Pope like Urban VII. He ordered Galileo to be tried by the Inquisition. Galileo was found guilty, forced to retract and was kept under house arrest for the rest of his life.

This is a tragic story, both for Galileo the man and for science. The effect was to put a total halt to science in the Mediterranean, the scientific revolution moving to Northern Europe. From a personal point of view our sympathies are with Galileo, although it must be said that he did not handle the affair as sensitively as he might have done. More broadly, the tragedy is that the established church of the time showed itself to have a closed mind to the truth when its authority was challenged. We are still experiencing the backlash today.

You will remember the key issue was whether the earth–the home of man, God's highest creation–is at the centre of the universe. It is interesting to ask how this is viewed by scientists today, towards the end of the twentieth century. We live in the post Einstein era. In 1905 Einstein published his special theory of relativity. In this he showed that there is no absolute frame of reference which enables one to identify the 'centre' of the universe. Any point travelling at any speed you care to choose travelling in any direction you care to choose is as good a 'centre' as any other. From the point of view of physics, the earth seems not to be a very special place. This in no way says that God does not view his creation, man, as special. This is an entirely different issue.

In recent years scientists have come to recognise that the very existence of man, a highly developed, intelligent being, able to communicate, says something about the nature of the universe in which he lives. This has become called the Anthropic Principle. As a simple example of this, one can assert that the universe must be so built that at least one planet has existed for sufficient time for man to have evolved. In this sense the scientist sees man as having a central role in the universe. We shall return to this issue in Chapter 3.

As regular as clockwork?

We have just considered the first major confrontation between science and the established church. The particular issue involved–is the earth at the centre of the solar system– does not seem to us in our generation to be a particularly significant one. But the next issue arising from it certainly is, and it has been one of the major determinants of Western culture for the last three hundred years. The central figure in all of this was possibly the greatest scientist of all time, Isaac Newton.

Newton was born at Woolsthorpe in Lincolnshire in 1642, the same year that Galileo died. His father had died before he was born and his mother remarried and left him to be brought up by his grandmother. Newton was a solitary, lonely man all his life. He never married and was always suspicious of people. He seems to have made little lasting impact either at school or as an undergraduate at Cambridge, but the two years after graduation, 1665 and 1666, were extraordinarily fruitful. He had intended to stay on at Cambridge, but they were the years of the Great Plague and, to escape it, Newton returned to Woolsthorpe. In just these two solitary years he established the fundamentals of the physics of his day, which remained unchallenged for two hundred years. It took him many years after he returned to Cambridge to fill in the detail and he was often reluctant to publish his findings. But all his great thought was conceived 'in the two plague years of 1665 and 1666, for in those days I was in the prime of my age for invention'.

Newton made major contributions to optics, to mathematics (he invented Fluxions, now known as Calculus), but above all to the understanding of the motion of all bodies including heavenly bodies.

In his two years at Woolsthorpe, Newton considered the motion of the moon around the earth. Is it just like a stone (or apple) that has been thrown so fast that although it is always being pulled towards the earth by gravity, it always missed it and follows a circular or elliptical path? Newton found that if the force of gravity on the moon weakened as the inverse square of the distance from the earth, then everything fitted. He correctly calculated the time taken for the moon to orbit around the earth. Later on he published a complete theory in the *Principia (Philosophiae Naturalis Principia Mathematica)*. By assuming the law governing the force of gravity was the *same* on earth and all over the universe, Newton was able to account for the motion of all the known planets and their moons, Kepler's laws and the trajectory of comets. He estimated the mass of the earth to within 10% of the currently accepted value. He calculated the mass of the sun and the planets. He predicted that the earth would be flattened at the poles and that the force of gravity would vary over the earth's surface. Newton had set the framework of physics that would remain unchallenged for two centuries.

The concept that the same type of force, gravity, was responsible for causing apples to fall on earth, and equally for keeping the earth and the planets in orbit around the sun, was revolutionary to people who were used to thinking of science in terms of *purpose*. From the time of ancient Greek physics it had been considered that the workings of nature were manifestations of the divine purpose. They would have said that an apple falls to the ground so that its seeds can germinate and produce more apple trees or so that we may enjoy 'the kindly fruits of the earth in their seasons'. They would also have said that the planets moved in a regular pattern to show the perfection of God's will. After Newton this framework was considered obsolete. It was supposed that the whole universe is governed by cause and effect. Purpose is irrelevant. The only cause is gravity making the planets to move as they do, or the apple to fall as it does. After Newton people asked the question *how* nature is as it is, not *why*.

Over the next hundred years or so, this way of thinking developed to cover the whole of life. The intellectuals of the eighteenth century called that period the Enlightenment or the Age of Reason. The attitude that everything could be understood by a logical, often mathematical, approach began to take hold of the thinking of the time and is still the foundation stone of our Western culture today. It has served us well in our material advancement, but the abandonment of an absolute purpose has resulted in unsatisfactory philosophies and spiritual poverty.

What happened to our view of God in all of this? For many, God seemed, and even today seems, irrelevant. The story goes that when the great French mathematician Pierre de Laplace discussed his monumental work on Newtonian dynamics *Systeme du Monde* with Napoleon Bonaparte, he was asked where God fitted into it. He gave the now famous reply, 'I had no need of that hypothesis'.

Many who still believed in God came to believe that his powers were limited by the laws of physics. He may have created the universe and got it going but now it runs like clockwork on its own. This view of God is called Deism and is very different from the Christian view of a loving, saving, almighty God who rules the universe.

One Christian response at the time was to suppose God was indeed constrained by the scientific laws that applied in areas that the scientific community understood. But it was possible to point to many areas where, in the eighteenth century, science had made little progress and God was still in control. What about the weather, medicine, the wonders of his living creation which were hardly understood at all? Thus was born the concept of the 'God of the gaps'. In this view, as time has moved on and our knowledge of science has advanced, the area left for God has become steadily smaller. Moreover, the Christians attempting to defend their position against the claim of an ever-advancing science, have inevitably appeared reactionary, anti-science and outdated, clinging desperately to an old-fashioned view that was no longer relevant.

The extreme rationalist view was put by Laplace:

> An intelligence which knew at one moment of time all the forces by which nature is animated and the respective positions of the entities which compose it, if moreover, this intelligence were vast enough to submit these data to analysis, it would embrace in the same formula the movements of the largest bodies in the Universe and those of the lightest atoms; nothing would be uncertain for it, and the future, like the past, would be present to its eyes.

In other words the future is completely determined by what is happening today. There is no room for free will, no room for God to act at all.

God in control

These deterministic views presented a serious intellectual challenge to Christianity. At first sight it might seem that the nineteenth-century development of electromagnetic wave theory by Maxwell provides a way out. But closer examination shows that this theory is still deterministic. Although, for reasons we shall discuss, determinism is no longer tenable in the light of twentieth-century science, it is still a view which is widely held. For this reason we will examine some of the important issues in the debate.

1) Underpinning it all, for the Christian, is the recognition that the world is God's world and he is in control. This applies to science just as much as anything else. Science is

never static. There are always new discoveries and new insights. This means that there will always be uncomfortable areas where, at a given point in time, scientific and Christian perceptions may appear to be in conflict. If this is the case Christians need to accept the difficulty but *never* regard it is a 'no go area' for God. Too often in the past, Christians have chosen the easy option of 'The God of the gaps'.

2) We now know that Laplace's contention that perfect knowledge of the current positions and velocities of atoms would enable an accurate prediction of their future positions to be made is simply wrong. This is discussed further below and in Appendix 1.

3) We now recognise that the orbiting of the planets around the sun is a rather special case where it is possible to calculate their motion very precisely. This is not the case with most physical systems. For complex systems, interesting patterns of behaviour are being observed which could not have been readily predicted on the basis of Newtonian mechanics.

4) Most scientists would accept that complex biological, psychological and sociological systems cannot be explained in terms of physics alone.

All this makes the deterministic arguments of the Enlightenment outdated and irrelevant. The modern scientific view of the world is much more complex and subtle. It is to these aspects that we must now turn. Inevitably in a book of this nature, it is only possible to skate over the surface of modern scientific developments. For ease of reading, only the basic concepts that are relevant to our discussion are presented in non-technical terms. Some of the areas are treated more fully, but still superficially, in the Appendices.

Nothing buttery

It is possible to set down a hierarchy of scientific disciplines so that the lower ones include the higher ones. A typical example would be:

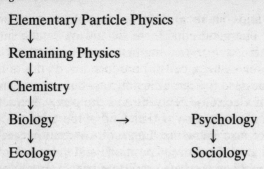

In this scheme we find there is nothing in chemistry that is contrary to the laws of physics, or nothing in biology where chemistry breaks down. The patterns of behaviour that were discovered in the higher, simpler disciplines, apply in the lower, more complex ones. This has led some people to go further and claim, for example, that biology is nothing but physics. This error has become known colloquially as 'nothing buttery'–rather more graphic than its formal name 'ontological reductionism'. Let us look at this in more detail.

Elementary particle physics is concerned with the basic building blocks of the universe. Until this century, scientists considered the basic building blocks were atoms but, in 1911, Rutherford showed that the atoms themselves had a structure with most of the mass being concentrated in a very small positively charged nucleus surrounded by a negatively charged cloud of light particles known as electrons. If atoms are very small, the nucleus is minute. If we were to magnify an atom to the size of a football pitch, the nucleus would be no larger than a coarse grain of sand. The next quarter of a century after Rutherford's discovery was probably the most exciting in the history of physics. By the end of it, it was well established that the nucleus itself was composed of building blocks of positively charged protons and neutral neutrons. Nuclei differed only in the number and proportion of these 'building blocks'. This was the basic physics that was exploited in the development of nuclear weapons and nuclear energy.

One finds out about the properties of a nucleus by shooting very energetic particles at it and observing how they bounce off. As time has progressed more and more powerful machines have been built to produce more and more energetic particles. By the 1950s it was clear that the simple picture of electrons, neutrons and protons was inadequate. All sorts of elementary particles were being found as larger and larger machines were built to fire increasingly energetic particles at the nucleus. Eventually a theory has emerged which again has produced a simple and elegant picture. In essence protons and neutrons are themselves made up of three types of building blocks known as quarks.

Now it would be true, according to our current perception of reality, that all matter is composed of quarks and some other elementary particles. But it would be wrong to say that, if we understand about quarks, we understand the whole of science. There is much more to it than that. In physics there are topics such as electricity, magnetism, why solids are hard, why metals conduct electricity whereas other materials are insulators, that you just cannot deduce from a knowledge of elementary particle physics alone. So although all matter is composed of fundamental particles, there is much more to physics than can be gleaned from a knowledge of elementary particle physics alone.

Similarly when one passes along the chain from physics to chemistry to biology to ecology, each new discipline has its own special contribution to make. Thus although living systems obey the laws of physics and chemistry, it would be wrong to say that biology was nothing but physics and chemistry. It is much more.

When one comes to study humans–how we think, how we behave in society, our self-awareness, our sense of justice and of right and wrong–we are in a different dimension once again. Humankind obeys the laws of biology, chemistry and physics but it is clearly wrong to say there is nothing more to us than that.

This point is stressed for the following reason. Many of

the advances in sciences such as biology have come from taking one aspect of the subject and analysing it in detail using physics or chemistry. The understanding of the genetic code in terms of the chemical structure of DNA is a well-known example. This method is a perfectly proper and powerful scientific approach. But, because of its success, some have been tempted to come to the illogical conclusion that, for example, biology is nothing but physics and chemistry.

The Taj Mahal is more than a heap of stone blocks. A Shakespeare sonnet is more than a collection of words on a sheet of paper. A man can be regarded as a collection of chemicals but there is more to it than that. For a Christian he is a precious creature, made in the image of God, given free will and possessing the potential to know, love and serve God.

Even more, God is above his creation. He cannot be reduced to mere science. He may choose to work normally through scientific laws, but to say he can be nothing more than that is clearly a nonsense.

A complex world

The picture of the universe envisaged by Newton and Laplace was essentially very simple. But the motion of the planets that they considered was a very special case. Most physical systems are more complex than that. In the next three sections we look at some of the phenomena which apply and which were completely unknown to the early scientists.

The first aspect we must consider is that any system left on its own will become more disordered—that is, arranged in a more random manner. For example, if one has a bucket of hot water and a bucket of cold water there is a degree of order—ie, hot water is in one place and cold water in another. However, if one mixes the two together one ends up with two buckets of tepid water. The universe has become

Nothing but a heap of stones?

Nothing but a collection of chemicals?

slightly more disordered because the order associated with the temperature difference has been lost. Moreover, one cannot readily reverse the process by starting from two buckets of tepid water and automatically ending up with separate buckets of hot and cold water; one has to do work to achieve this. This is known as the Second Law of Thermodynamics.

Another example that is often used to illustrate this concept is that of the snooker balls. At the beginning of the game the red balls are all put together in a triangular pattern. They are thus in a highly ordered arrangement, but one that can easily be disrupted if the white ball is directed at it. If this was filmed, it would be possible to judge whether the film was being run backwards or forwards by seeing if the red balls were flying apart all over the table when struck, or whether they were coming together, as if by magic, into the ordered triangular pattern. Our intuition tells us immediately which is the correct way to run the film: as time progresses the balls fly apart–ie, the arrangement becomes more disordered.

What is true of the snooker balls or buckets of water is true of the universe as a whole. It is gradually running down and will eventually degrade, albeit on a timescale of many billions of years. There can be localised regions of the universe where the order actually increases. A living organism is just such a place. But this concentration of order is more than offset by an increase in disorder elsewhere.

There are important consequences from all of this. To some people, the universe seems so large and permanent that they find it hard to envisage God creating it, and even more God bringing things to an end at some point in the future. And yet the Second Law of Thermodynamics itself points to a beginning and an end. If the universe is becoming less ordered with time, if we were to go back in time it would become more and more ordered, and, if we were to go back sufficiently far it would be completely ordered. We cannot then go back even further in time since it would require still

more order. Hence there must have been a beginning to the
universe as we know it. It has not been there for ever. This
is, of course, completely consistent with the Christian view
of God creating the Heavens and the Earth. Secondly, the
universe will not support life for ever. There is an end to life
on earth–but not of course to our spiritual life which is
eternal in both quality and time. Again this is consistent with
the Christian view of the world as a temporary home to be
lived in until God's final judgement on mankind at some
point in the future.

As the universe becomes more disordered with time,
some of the apparently simple parts-of it become more
complex. In recent years, scientists have come to recognise
that the detailed behaviour of most complex systems is
impossible to determine, even though the deterministic laws
governing this behaviour still apply. We now turn to consider
this in more detail.

Chaos and order

Returning for a moment to Laplace, we will recall that he
considered that if you knew the state of the universe at one
moment in time, then you could calculate its state (at least in
principle) at all earlier and later times. We have seen that
this may be a good approximation for planetary systems but,
in recent decades, it has become increasingly apparent that
most physical systems are so complex that even with powerful computers such calculations are impossible in practice.

The basic reason for this is that with many systems it only
requires a small change in the initial parameters to result in a
major change in the behaviour. An example of this is the
difficulty of medium-term weather forecasting. It is possible
to write down simple mathematical equations that describe
the weather at one instant in time in terms of the weather a
short time previously. These can then be solved numerically
using a computer. By repeating the calculations many times
one can build up a forecast for several days ahead. In prac-

tice an extremely powerful computer is required to calculate the results faster than real time.

But although one has many weather observations to provide the starting information to feed into the computer, these cannot be perfectly accurate. It is found that these small inaccuracies tend to lead to ever-increasing errors in the predictions of future weather as the length of the prediction increases. This limits the potential of medium-range weather forecasting to a few days ahead. It is not that the computer is wrong, it is not that the equations being solved are wrong, it is simply that small errors fed into the input magnify as time goes on. This effect was first discovered in 1963 by Edward Lorenz. He called it the Butterfly Effect. He made his point dramatically by suggesting that the flapping of a single butterfly's wing today in one part of the world would produce a tiny change in the state of the atmosphere. The effects could then build up so that, some time in the future, a hurricane would strike at some place in the world other than that where it would have done had there been no butterfly.

One can illustrate the difficulty of predicting the future with an even simpler example – a ball-point pen on a table. If you lay the pen on the table and cause it to roll, then one can readily predict its motion. It will continue to roll until its rotational energy has been dissipated by friction (a degredation mechanism). Now stand the pen vertically on its point and try to predict what will happen. Clearly it will fall over but it is impossible to predict the direction. If one has stood it up extremely accurately it will take longer to fall but, eventually, some draught of air or a vibration will set it moving and it will topple. One could elaborate the system – mount it in a vacuum chamber and on a shock-absorbing platform – but eventually it will fall in an unpredictable direction.

In recent years scientists have been directing increasing attention to those systems that obey precise deterministic laws so that, although in principle their behaviour can be calculated, in practice it cannot. The subject, which has become fashionable over the last decade, is called Chaos. It is found that in this chaotic behaviour, patterns often emerge which are not related in a simple and straightforward manner to the underlying laws.

Consider, for example, a snowflake. This is just an ice crystal, but of a particularly beautiful form. The snowflake grows when water molecules condense on it in the location that is easiest for them to do so. The physics is simple. But who, without seeing one, would have predicted from first principles that such a beautiful and complex form could arise, let alone the precise shape?

This is just one simple example. Further details of aspects of Chaos theory are given in Appendix 2. But for our present purposes it is just important to note that nature is complex. Although it may be possible to enumerate simple basic principles that a physical system obeys, in practice it is often impossible to calculate the behaviour of that system. In particular the systems can behave in a complex, sometimes chaotic manner. And yet out of this complexity there often emerges a new kind of order that would have been difficult to have predicted *ab initio*.

All this throws serious doubt on the claim that the future can be accurately predicted from a knowledge of the present, particularly when applied to systems as complex as living organisms. These doubts are increased when we come to consider the role of chance, the subject of the next section.

Blind chance?

The behaviour that has been described so far in this chapter is complex–but it is deterministic. In other words, if it were possible to repeat an experiment exactly, you would get the same results every time.

The startling discovery of the 1920s was that nature is not like that. And the departure from determinism is most marked for very small particles. Newton's laws of motion are very good approximations to reality when one is dealing with normal-sized or large objects. They break down completely at the level of the atom and below. On this scale it is necessary to use quantum theory. This is described in Appendix 3. But, for the purposes of this chapter, it is only necessary to understand one simple quantum theory concept known as Heisenberg's Uncertainty Principle. This states that it is impossible to know exactly both the speed and the position of a particle—and the lighter the particle, the greater the uncertainty. There is a sort of graininess in nature and there is no way of overcoming this. If you know where a particle is, you cannot say precisely what it is doing. If you know how fast a particle is moving, you cannot say precisely where it is.

Quantum theory is one of the foundations—probably the greatest foundation—of modern physics. It is counter intuitive. Even Einstein found it hard to accept, believing that if one looked sufficiently hard one would find a deeper, possibly deterministic, reality within the randomness. In a famous letter to Niels Bohr, he asks the question, 'Does God play dice?' But today the vast majority of physicists would accept that, although some fundamental uncertainties remain, quantum theory is valid and that there is a real breakdown of Newtonian determinism. There is a degree of randomness built into the laws of nature.

Some would go further and claim that biology is dominated by chance. Jaques Monod, in his famous book on molecular biology *Chance and Necessity*, wrote:

> Pure chance, absolutely free but blind, is at the very root of the stupendous edifice of evolution. . . .

> Man at last knows that he is alone in the unfeeling immensity of the universe, out of which he emerged by chance. Neither his destiny nor his duty have been written down.

What a change from Newton and Laplace! The determinism which gave us no room for independent action has gone, to be replaced by blind chance, making us a meaningless accident.

The Christian would not, of course, accept these views. What would he say to all of this?

In the first place he would recognise that the laws of nature contain a very subtle interplay between randomness and determinism. In devising these laws God has trodden a path that avoids both the inflexibility of the Newtonian clockwork universe and the irrationality of the completely random one, while gaining the benefits of the fruitfulness of the two together. We have seen how, in the example of the snowflake, randomness and complex determinism combine to provide a remarkable structure. In Chapter 4 we will consider life itself where, contrary to Monod's despairing view of the futility of the world, we shall be claiming that the combination of randomness and the determinism was God's way to allow life to evolve, culminating eventually in the emergence of man.

Secondly, we note that the universe in which we live is capable of giving us surprises. Newton's work was outstanding by any standard. But he would have been astonished at the developments that have taken place in the last three hundred years, casting an entirely different light on the universe he studied. Science is always provisional–today's science being merely the best we can do at the present time.

I often wonder what Christians living at the height of the Enlightenment made of Newtonian determinism. I hope they were able to stand back and see that they must be more than mere automatons. If this is all they were, how could they know this, or indeed anything in a meaningful way? In our own age, if we are to claim to be capable of sound judgement and understanding, then surely we must be more than the product of blind chance.

So thirdly, the Christian would take issue with Monod

and claim that man's destiny and his duty have indeed been 'written down'. It is not surprising that Monod comes to believe in a world without purpose if God is left out of the equation.

3

In the Beginning

'When I look at thy heavens, the work of thy fingers,
the moon and the stars which thou hast established;
what is man that thou art mindful of him,
And the son of man that thou dost care for him?'

(Ps 8:3-4).

How great thou art

IN THIS CHAPTER we are going to look at the universe, its size and its purpose. Psalm 8, which was probably composed nearly three thousand years ago, reflects the recognition of the great wonder of the universe and man's apparent insignificance in it. It is a feeling that many would share today. The psalmist would have been used to looking at the stars under a clear desert sky. He could not fail to be impressed by the number he could see, But this is nothing compared with the appreciation we have today at the vastness of the universe, its complexity and the precision with which it is made.

The psalmist (David?) goes on to say of man:

Yet thou hast made him little less than God,
and dost crown him with glory and honour.
Thou hast given him dominion over
the works of thy hands;
thou hast put all things under his feet (Ps 8:5-6).

In spite of the magnificence and vastness of the whole of God's creation, man can still regard himself as very special.

Modern science does not disagree with this view. Indeed it has become clear in the last few decades that the universe is very finely tuned and that if the relationship between the basic physical constants differed only slightly from their observed values, then life as we know it could not have evolved and man could not exist. The psalmist could recognise by faith that man is made in the image of God and is important to God in spite of the vastness of the universe in which he occupies only a minute corner. We too share that faith, but it is strengthened by our scientific understanding of the very special nature of the physical universe in which we live, a universe which seems to have been carefully designed for us to inhabit.

The modern scientific view of how the universe began and how it operates is remarkable. Surprisingly, in spite of a number of television programmes and popular books on the topic, it is not particularly well known. The essence of the story is that our world is made up of the debris of stars that existed earlier in the history of the universe. These early stars were essentially nuclear reactors, building up chemicals such as carbon that are indispensible for life as we know it. Some of these stars eventually exploded, scattering the debris into space. This process itself took a few billion years. Eventually some of the debris collected to form the earth. It took a further four and a half billion years for humans to develop. So the universe must have existed for a long time in order for humans to emerge. We shall see that only a very special universe could be sufficiently long lived for this process to have taken place. In particular it must be vast and contain a very specific average density of matter, much of which has condensed into stars. So if it were not for the myriad of stars, which so impressed the psalmist, we could not exist. What a remarkable story!

The Christian can rejoice in the great Creator who, billions of years ago, planned such an amazing world for us to inhabit. Even the scientist who is not a believer normally feels a sense of awe at the wonder of the universe. In the next few sections we will look at this in more detail.

The life history of a star

When scientists examine the light emitted by a star, they can obtain information on the relative abundance of the chemical elements in it. A standard picture emerges. Most of the star consists of hydrogen. Between 23 and 25% of the mass of the star is helium. Then there are small amounts of the remaining 90 chemical elements – never exceeding 2% of the total and often less.

A star is created from the large volume of hydrogen and helium gas that is circulating in the universe. If one part of this gas is slightly more dense than the rest, the additional gravitational attraction begins to pull matter towards this denser area, the density of which is thereby further increased. Eventually the gas collapses into a massive ball. The gravitational energy is converted into heat energy. This has two consequences. One is that the pressure builds up in the ball, preventing further gravitational collapse. The second is that, with the high temperatures at the centre of the ball, thermonuclear reactions take place with the release of large amounts of energy. This energy is sufficient to keep the star hot and the pressure high even though it is losing energy from its surface in the form of heat and visible radiation. The thermonuclear reaction is the same one that occurs in a hydrogen bomb. The nuclei of the hydrogen atoms are combining to form helium with the release of energy.

At the same time helium nuclei are combining to form heavier nuclei. This process continues, building up heavier and heavier nuclei until iron is formed. Iron is the most tightly bound nucleus. You have to supply energy to make elements heavier than iron and these are therefore much less abundant than the lighter ones. This process of nucleosynthesis continues until the hydrogen fuel runs out.

The sun is a typical star. It has been shining for about four and a half billion years and it is believed that eventually, in a further five billion years, the hydrogen in its core will run out. It will then swell up to become a red giant before settling down as a white dwarf–less than 1% of its present size.

If, however, a star has more than 1.4 times[1] the mass of the sun, the gravitational attraction is sufficiently great for the star to collapse at the end of its life to a very high density, comparable to that of the atomic nucleus. This formation of a neutron star with a diameter of about 10 miles is accompanied by a massive explosion and ejection of the outer regions of the star into space. This is a supernova where the brightness of the star increases perhaps a billion fold for a few weeks. During the explosion, conditions are sufficiently hot to produce quantities of nuclei that are heavier than iron. So, as a result of the supernova explosion, the material in space, which was originally essentially just hydrogen and helium, has become enriched with the other elements which were created inside the star.

It is this enriched hydrogen and helium that now collapses under gravity to form a second generation star. If, like our sun, there are planets associated with it, these planets will also contain this mix of atoms. So it is with our earth. Most of the hydrogen and helium has escaped–since they are light the relatively weak earth's gravitational field is insufficient to hold them. But it is believed that almost every atom other than hydrogen and helium that makes up the earth or your or my body was created, billions of years ago, in a star that is now dead.

This fact is astonishing in itself. But equally remarkable is the fact that the process of nucleosynthesis could not have proceeded beyond helium if the properties of the nuclei had been only very slightly different. This is just one example of the remarkable fine tuning that is built into the physics of the universe.

Carbon is made by the fusing together of three helium

nuclei in the high temperature and pressure conditions that prevail in the interior of a star. It is highly unlikely that three nuclei will come close together simultaneously, sufficiently frequently for a large quantity of carbon to be produced. However, two helium nuclei can form beryllium which lasts long enough (only 10^{-17} sec $- 10^{-17}$ sec means that one with seventeen noughts of these minute intervals of time would be required to make up a second) for it to collide with a helium nucleus and form carbon. There are specific energy levels (resonance levels) in nuclei where reaction rates are very much higher than normal. When beryllium and helium combine they release 7.37 MeV of energy. (Don't worry about MeV – it is just a unit of energy.) For a sufficient reaction rate this, plus the thermal energy they had, must exactly match the resonance energy in carbon which is 7.65 MeV. That they do just that is a remarkable 'coincidence'.

But this is not the end of the story. If it were possible for the carbon nucleus to react resonantly with the helium nuclei, then the carbon would be destroyed and oxygen formed. There would be no carbon-based life as we know it. But this time the oxygen resonance level (7.12 MeV) is just *below* that of carbon plus helium (7.16 MeV), so this time the thermal energy cannot cause the necessary coincidence for a rapid reaction rate and the carbon is not immediately destroyed.

This is just one example of the many remarkable 'coincidences' that exist. Sir Fred Hoyle, who led much of the early work on nucleosynthesis in stars, and who would not claim to be a Christian, wrote: 'I do not believe that any scientist who examined the evidence would fail to draw the inference that the laws of nuclear physics have been deliberately designed with regard to the consequences they produce inside the stars.' The Christian would not disagree with this and would be able to point to the Great Designer. Truly we live in a 'Designer Universe'.

The universe as a whole

We have looked at the life history of a star, a building block of the universe. It is now clear that stars are grouped in space to form galaxies. The sun is a typical star in a typical galaxy we know as the Milky Way. The Milky Way consists of some 10^{11} stars and is about a hundred thousand light years in scale. It is known as a disc galaxy, the stars moving in nearly circular orbits. They are prevented from flying apart by their mutual gravitational attraction. In some galaxies, known as elliptical galaxies, the stars are swarming around in more random directions.

The galaxies themselves are not distributed at random but tend to cluster in groups of perhaps one hundred thousand, with a scale of seveal million lights years across. The structure of these clusters is not fully understood but is probably associated with perturbations in the density of the early universe. So although the distribution of galaxies in the universe is, on average, fairly uniform, when one looks in detail, a complex structure emerges.

The distant parts of the universe we observe is the result of radiation that set off millions of years ago from hot bodies such as stars. One can study the motion of stars orbiting in a galaxy, or of galaxies in a cluster, by measuring the relative velocities of the stars or galaxies spectroscopically. It is then possible to calculate the balance between the gravitational and centrifugal forces. It is found that they do not balance each other unless there is some additional matter there beyond that which can be seen. We know the stars are there because we can see them. It seems that most of the material in the universe is invisible–the luminous stars comprising only 10% or less of the total. This conclusion is confirmed when we examine the motion of the universe as a whole.

When we look at the universe we find that it is truly massive. We can get an idea from the following multiplication table:

By mass

1 earth	=	10^{23} men.
1 sun	=	3×10^5 earths.
1 galaxy	=	10^{11} suns.
Universe	=	10^{10} galaxies.
	=	10^{21} stars.

(Here 10^{11} means the number 1 with eleven zeros after it.)

It is difficult to imagine the significance of numbers such as this. To give some feel as to what it means, let us represent each star by a pea and let us first ask how large a pile would be made by 10^{11} peas–the number of stars in our galaxy. The answer is that if they were spread around a typical cricket ground they would reach a height of three feet.

Now let us consider the stars in the whole universe. If we take our 10^{21} peas and spread them over the whole of Great Britain, they would make a pile a quarter of a mile high!

This is remarkable and indicates the vastness of the universe. It is the scale on which God the Creator works.

We would of course be wrong to imagine our universe as being tightly packed like a pile of peas. There are vast distances between the stars. There is, on average, only one atom in every cubic metre of the universe and, with the possible exception of the dark matter, this is mainly concentrated in certain regions such as the stars. Once again, if we were to scale down the stars each to the size of a pea, the average separation between the peas would be a hundred thousand kilometres–a quarter of the way to the moon.

So it is clear that the universe is vast. We must now look to see how scientists consider it could have begun, some ten to twenty billion years ago.

The Big Bang

In the last thirty years, a scientific consensus has emerged that it is highly probable that our universe started in a massive explosion known as the Hot Big Bang. Any theory must account for the following basic facts about the Universe:

1) It is massive.

2) It is expanding.

3) The visible matter consists mainly of hydrogen and helium in the ratio three to one by mass. Heavier elements also exist and can be accounted for by nucleosynthesis in stars, as already described.

4) On average the universe seems very uniform but the visible matter is concentrated in stars which are clustered in galaxies which themselves tend to cluster.

5) There are billions of times more particles of electro-magnetic radiation (photons) in the universe than particles of matter.

6) Most of this radiation is concentrated in very long waves (microwaves). We are used to seeing visible radiation from hot surfaces (we talk about a very hot piece of metal as being red hot). This microwave radiation is typical of that from a surface at only three degrees above absolute zero (minus 270°c). it is also remarkably uniform. After making due allowance for the motion of the earth around the centre of the Milky Way, it is the same in all directions to at least a few parts in one hundred thousand. It was only in 1992 that the small fluctuations in the background were discovered by the COBE satellite.

In considering the Big Bang theory, it is convenient to start to describe this model of the universe starting about one second after the initiation. By this time the density had fallen to around that of a normal solid or liquid and the temperature to around ten billion degrees. We have reached a regime where our understanding of the physics is on a very firm, well-tested foundation. In fact we have reached the point where, in many ways, the universe looked like an enormous hydrogen bomb explosion.

If we were able to look at the composition of the universe at that point in time, we would have found that most of the energy resided in photons—billions of photons for every neutron and proton present. The nuclear reactions taking place would maintain an equal number of protons and neutrons at this point in time but, over the next ten minutes, as the temperature fell to half a billion degrees, the number of protons would have increased somewhat at the expense of the neutrons, which would then have reacted with protons to form helium nuclei. Careful calculations of this process have been carried out. It is predicted that the reactions should have resulted in a proton to helium mass ratio of three to one, the ratio we observe in stars today. The success in predicting this ratio is one of the powerful arguments in favour of the Hot Big Bang theory.

A second argument relates to the microwave radiation. When the process of forming the elements hydrogen (protons) and helium was completed after ten minutes, the temperature was half a million degrees. Under these conditions the electromagnetic radiation (mainly powerful X-rays at these temperatures) was strongly scattered by the matter present and the two were thereby closely coupled. As the universe expanded, its temperature fell until after three hundred thousand years a figure of around four thousand degrees was reached. At this level the hydrogen and helium ceased to be ionised and the universe became transparent. The microwave electromagnetic radiation we now see is that which set off at this time from a remote part of the universe and is reaching us now. Because it has travelled so far and for so long it is 'red shifted' down to the microwave region. This picture is confirmed by the detailed calculations that have been carried out.

The Hot Big Bang theory as we have described it after the first second therefore accounts very well for the facts 1), 2), 3) and 6) enumerated above. It says nothing about 4) and 5) and also raises a major further question of its own. This is how conditions were just right for the universe to be at a 'critical density'. Let me explain. As a result of the Big Bang, matter was projected at enormous speed and has been slowing down under the force of gravity ever since the first fraction of a second. If the initial speed had been only very slightly less than it was, the universe would have collapsed long ago. If it had been only very slightly greater than it was, gravity would not have slowed it down sufficiently for the universe to be dense enough for stars and galaxies to form. There would have been no chemical elements other than hydrogen and helium. There would be no life. This is why there are so many stars. There had to be such a vast number to provide the gravitational attraction to prevent the universe from flying apart. The fine tuning had to be very precise in the early stages of the universe for the density to be sufficiently right today. At one second after the start of the Big Bang, conditions had to be precise to one part in 10^{15} – an unimaginable accuracy.

So we are left with four questions:

1) Why are there billions of photons for every atom?

2) Why is the universe so homogeneous?

3) Can we account for the density perturbations that led to galaxies and stars?

4) Why is the density so close to the critical density?

In order to seek an answer to these questions, scientists are trying to examine what could have happened in the first second of creation. We will discuss this now.

The first second

Astronomers become accustomed to and get a feel for the vastness of the universe. But it seems that many of the most fundamental questions relate to the time when the universe was small, dense and extremely hot. Such conditions are those that are normally the province, not of astronomy, but of high energy particle physics. Over the last fifteen years or so there has been an increasingly fruitful collaboration between cosmologists and elementary particle physicists. It is now generally accepted that there were two times in the early history of the universe that were highly significance. Both are so early and the conditions so extreme that it is hard to visualise them. One critical time was 10^{-35} of a second (yes, one divided by one and thirty-five noughts of a second!). At this time the temperature of the universe was 10^{27} degrees and the energy of any particle was 10^{14} GeV. (A GeV is the energy you get on a particle with a charge equal to that of a single electron when accelerated by an electric potential of a billion volts.) Clearly these are the most extreme conditions. The energies involved are a hundred billion times more than have been produced on earth in the most powerful particle accelerators. Nevertheless, in the last ten or fifteen years, theoretical physicists have come to believe that these conditions are important because they define a point where three of the fundamental forces in nature become unified. Let me explain.

Two hundred years ago, it was considered that electricity and magnetism were two separate, unrelated entities. There is a magnetic force which is manifested when you pick up iron with a magnet and there is also an electric force by which you can attract small particles to an insulator after rubbing and thereby electrifying it. But during the next hundred years it became clear that these two forces were interrelated. You can produce a magnetic field by passing an electric current through a coil of wire and also you can generate an electric field by varying a magnetic field. So although it may appear that there are separate electric forces and magnetic forces, in fact they are separate aspects of a single electromagnetic force.

It is considered that there are four fundamental forces in nature. The most powerful is appropriately known as the strong force. It acts over a distance comparable with the size of a nucleus and serves to bind the neutrons and protons inside the nucleus. Next in strength is the electromagnetic force which we have just described. Then there is the weak force which has a very short range and is significant in radioactive decay. Finally there is the weakest force of all on the elementary particle scale—the force of gravity. However, gravity acts over long distances and moreover, the gravitational fields from all the vast number of nuclei add up, making it the most significant force for massive bodies such as stars or galaxies.

It is now known that it is possible to combine the weak and electromagnetic forces into a single theoretical model. Above about 10^{15} degrees, or 100 GeV in energy terms, these two forces effectively merge into a single force, the electroweak force. Similarly there are theories that suggest that the strong force and the electroweak force could merge at temperatures above 10^{27} degrees–the conditions prevailing at our first significant time after the initiation of the big bang of 10^{-35} seconds.

It has been suggested that as the universe cooled through 10^{27} degrees, the strong and electroweak forces split into the two separate forces. However, there may have been a delay of perhaps 10^{-30} seconds before this so called 'spontaneous symmetry breaking' occurred. There are arguments to suggest that during this period, the universe might have expanded enormously and very rapidly–a phenomenon known as 'inflation'. Put rather crudely, the inflation stretched the early universe, smoothing any initial density fluctuations, and it has been suggested that as a result of this process the solutions to questions 2), 3) and 4) may possibly be found.

The answer to the first question–why are there so many photons for every particle of matter in the universe?–is somewhat different. It is due to a minute asymmetry between the behaviour of matter and antimatter and is discussed in Appendix 4. If this asymmetry did not exist there could be no matter in the universe—and no life.

We started this section by saying there were two critical times in the early history of the universe. We have indicated the importance of events near 10^{-35} second. The other time is even earlier–10^{-43} seconds after the initiation of the Big Bang. At this time–known as the Planck time–it is believed that gravity would have been merged with the other three forces into a single unified force. If there were to be satisfactory theories devised about the origin of the universe, it would be necessary first to develop an appropriate theory about the extreme conditions that existed at this early moment in time. This is a challenge that is being taken up by theoretical physicists around the world. A single theory dealing with all four forces is sometimes, rather ambitiously, called 'A Theory of Everything'.

There have been many interesting suggestions made about this very early universe. For example, could it have been initiated by a quantum fluctuation? We saw in Chapter 2 that, as a consequence of the Heisenberg Uncertainty Principle, it is not possible to specify precisely both the position and speed of a particle. Similarly, it is not possible to specify both energy and time. As a consequence, there are short blips of energy, even in a vacuum! In such an energy blip, a gravitational field would be set up because of the Einstein equivalence of mass and energy. The energy in a gravitational field is negative so more energy would be available to make the whole process take off. These concepts may seem strange, but the world of quantum physics is a strange world. This particular theory may not be correct, but it is mentioned to give a flavour of the ideas being pursued.

Another approach is a version of the Anthropic Principle. This, you will recall, says the nature of the universe must be consistent with man's existence in it. Now, we have seen that the laws of nature must have been finely tuned for such a universe to have developed. It has been suggested that perhaps there were a vast number of universes created–all of different sizes and with different physical properties. We just happen to be in the one that got the conditions just right for life to develop.

There has been much publicity given to the ideas of Stephen Hawking, both as a result of his best-selling book *A Brief History of Time*, and television programmes about him and his work. Because of this widespread interest, the issues raised are discussed in Appendix 5.

Many Christians find the ideas we have been discussing

thought provoking—some find them disturbing. What can we say about this?

In the first place, science can only go back to some limit in time. It may be possible to postulate a mechanism that might have led to the formation of the material universe at this early stage, but it can say nothing about the origin of the laws of nature, which must already have been in place for the postulated scientific process to operate. Although science can explore the ground rules within which it operates, it is powerless to discuss why they are as they are. God's creative acts must have preceeded what, with our scientific eyes, we perceive as the limit of time. One could say that the stage was set before the curtains were drawn and the play began. We see the play but not the preparation.

In the second place they are just ideas and are impossible to prove. It is a perfectly valid exercise to produce a mathematical model of the very early universe, but we must recognise it to be nothing more than that. There is no direct scientific evidence of the conditions prevailing at 10^{-35} second. The energies involved are a hundred billion times more powerful than the ones of which we have direct experience. The conditions at the Planck time (10^{-43} seconds) are even more extreme. This is not to belittle the brilliance of the scientists producing the theories. But the nature of the problem is such that it would be extremely difficult, if not impossible, to prove them.

But even if they were true, they in no way rule out God's involvement in the processes. Stephen Hawking recognises this point when, in the conclusion to his book, he writes:

> Even if there is only one possible unified theory, it is just a set of rules and equations. What is it that breathes fire into the equations and makes a universe for them to describe? The usual approach of science of constructing a mathematical model cannot answer the question of why there should be a universe for the model to describe.

As we point out so frequently in this book, the fact that we might understand some phenomenon scientifically, in no way excludes God from having designed it. To think otherwise is as silly as to think that no one could have designed a motor car because some people understand how it works.

No, what we have been given is a glimpse of our God's handiwork in all the wonder and complexity of his physical universe.

Note

1. If the initial star has more than about 2.5 times the mass of the sun, the force of gravity is so strong that the collapse continues beyond the neutron star stage to form a black hole. Here the gravitational forces are so great that nothing, not even light, can escape from its inner regions. Nevertheless, there is radiation from the outer regions of a black hole and the brightest objects in the distant galaxies–quasars–are believed to be associated with massive black holes.

4

Life on Earth

'Evolution is the most powerful and the most comprehensive idea that has ever arisen on earth.'

Julian Huxley.

'I see no good reasons why the views given in this volume should shock the religious feelings of anyone.'

Charles Darwin, *Origin of Species*.

Challenge to Genesis

'SURELY SCIENCE HAS DISPROVED the stories of creation and early human history that are presented in the early chapters of the Bible.' This is one of the most common challenges faced by Christians when the subject of science and religion is discussed. The implied argument is that if the Bible is wrong right at its beginning, then surely it is unreliable all the way through. Unfortunately Christians are divided in their response. Some would challenge the science on which the assertion is based. However, most would accept that modern science provides the best scientific understanding of how life has evolved, and would claim that the stories in the early chapters of Genesis were not meant to reflect the insights of late twentieth century science (some of which will inevitably be out of date in a hundred years' time) but were accounts of timeless spiritual truths expressed in language that would have been understood at the time the Old Testament was written. Before we examine these approaches in more detail, it is helpful to remind ourselves of the historical background to this debate.

Natural theology

From the time of Newton to the first half of the nineteenth century, there was considerable emphasis on natural theology. This sought to discern the nature of God and his purposes from science and nature. Great emphasis was laid on proving the existence of a creator from the beauty, order and complexity of his creation. The climax of natural theology came with the publication in 1802 of the book *Natural Theology—or Evidences of the Existence and Attributes of the Deity Collected from the Appearances of Nature* by William Paley. In this book Paley begins by arguing that if one was walking across a heath and kicked a stone that was lying there, one would find it hard to argue that such an object had not lain there for ever. However, if one found a watch it would be obvious that the watch had been designed and built by someone with a purpose in mind. If this is true of a watch, how much more true is this of nature which is so much more complex?

The pendulum has swung and natural theology has been out of fashion although there are signs of a revival in interest in recent years. There were a number of reasons for the discrediting of natural theology. The first is that one cannot push 'design arguments' too far. One cannot use them to prove the existence of God in a formal, logical, mathematical sense. As early as 1779, the philosopher David Hume demonstrated the weaknesses in such arguments. But even though the absolute deductive proofs fall down, Christians can legitimately take great comfort from the fact that a complex, ordered nature such as we see is entirely consistent with it having been created by our wonderful God. We may not be able to prove God from it—but having believed in God for other reasons, we can be greatly encouraged by the signs of his handiwork in nature. But beyond this, the weight of evidence from nature strongly supports the probability of God's existence.

The second reason why natural theology went out of favour was that it was used by Deists and Pantheists to support their narrow views of God as opposed to the full Christian views revealed in Scripture. You will recall that Deists believe that God created the universe and then left it alone to run its course. Pantheists believe that God and nature are identical. Christians believe that there are signs of God to be discerned in nature. For example, St Paul, in his Epistle to the Romans (chapter 1:18-20), explains that even those who have not heard the Christian Gospel have no excuse for their sinful behaviour because God is revealed in his creation:

> For the wrath of God is revealed from heaven against all ungodliness and wickedness of men who by their wickedness suppress the truth. For what can be known about God is plain to them, because God has shown it to them. Ever since the creation of the world his invisible nature, namely, his eternal power and deity, has been clearly perceived in the things that have been made.

But Paul was fully aware that God's revelation of himself is much deeper than that which can be discerned in nature alone. Just before this passage he had written:

> For I am not ashamed of the gospel: it is the power of God for salvation to every one who has faith, to the Jew first and also to the Greek. For in it the righteousness of God is revealed through faith for faith (Rom 1:16-17).

Christians were suspicious of natural theology because people were using it to the exclusion of God's fuller revelation of himself and his purposes through the Scriptures and the traditional teaching of the church. It is so easy for natural theology to be man centred rather than God centred.

But the greatest blow to natural theology came in 1859

when Darwin published *The Origin of Species*. Animals had evolved into their current species, he argued, through the forces of natural selection. In effect, he was saying that although it may look as though there is a purpose behind nature, the process of natural selection is sufficient to account for the complexity and diversity we see in it. But more of this later. We must first of all turn to an earlier debate about Noah's flood and geology.

The messages in the rocks

In this section we describe current scientific views on the development of the earth. In Chapter 3 it was explained that it is considered that the earth was formed from the debris of supernova explosions coming together by gravitational self-attraction. This debris was in the form of small bodies known as planetesimals made up of silicon compounds, iron and magnesium oxides and smaller amounts of all the natural chemical elements. Because a supernova is a thermonuclear explosion, much of this material was (and some still is) radioactive. We all live on top of a natural nuclear waste dump!

The earth was formed about 4.7 billion years ago. It heated up because the kinetic energy of the planetesimals was converted into heat energy on impact with the growing earth. Gravitational energy was also released as the earth compacted, and finally the young earth heated up further as the radioactivity in it slowly decayed. Within a billion years the iron on the earth began to melt, migrated to the centre displacing the lighter material upwards and releasing further gravitational energy in the process.

In this way, chemical compounds rose to the surface, cooled and formed the earth's crust. The earth's early atmosphere is thought to have consisted of water vapour, carbon dioxide and nitrogen which were released in this process (a process that still occurs today when volcanoes erupt). Much of the water vapour condensed to form the oceans.

It is possible to date the earliest rocks from measurements of their radioactivity. It is believed that they are about 4 billion years old. Much of this rock has been eroded, transported by water and reformed, often several times during the subsequent earth's history. Massive sections of the earth's crust are still moving slowly as a result of convection currents within the earth, mountain ranges being formed when two sections of the earth's crust collide.

Such is our modern view of the origin of the earth's crust. It has all developed from key studies undertaken early in the nineteenth century. Prior to that time many thought that the fossils buried in rocks and earth were the remains of animals that had been overwhelmed in the flood at the time of Noah described in Genesis.

This became known as the diluvian theory. According to this, the rocks were compacted soil that had been washed there by the flood. It was also believed that this catastrophe had occurred less than 6,000 years ago since Ussher and Lightfoot had concluded from the genealogies in Genesis that the world had been created in 4004 BC. This view was undermined by the evidence that was being assembled by men such as William Smith, a surveyor involved with the excavations associated with the many canals that were being constructed in the West of England. He recognised that each type of fossil was associated with a specific bed of rock. By 1830 the issue was clear. In that year Lyell published his famous book *Principles of Geology* and in the following one Adam Sedgwick, the Professor of Geology at Cambridge and a devout Christian, publicly announced that he had previously been wrong to support the diluvian theory. The processes that had led to the deposition of fossils in the sedimentary rocks were slow and had taken a very long time indeed. This view has been confirmed over the years, particularly with the development of techniques for dating the rocks by measurements of its radioactivity.

Sedgwick never doubted the historicity of Noah's flood. What was being challenged was the assumption that this flood was responsible for laying down all the fossils. It is always important to distinguish between what the Bible actually says and some interpretations of it which are of doubtful validity.

Sedgwick's contribution to the science of geology was enormous. Even so, in his busy life he found time to devote to his students. One of these, whom he took to North Wales on a fossil-hunting expedition in 1831, was a young man called Charles Darwin.

Evolution – God's solution

The story of Charles Darwin and the voyage of 'The Beagle' is well known. In 1831, at the age of twenty-two, Darwin joined the ship as a cabin companion to the Captain, Robert Fitzroy. He was selected because of his knowledge of natural history and one of his tasks was to gather scientific data. The voyage was to last five years and to expose Darwin to the great variety of living creatures on the islands off the South American coast. The fossils on these islands and the mainland were also remarkable, covering a wide range of living and extinct species. Clearly creation was more complex than Darwin had hitherto envisaged. Why did each of the Galapagos Islands have its own characteristic species of tortoises, mocking birds and finches? Had God created all these separate species from the outset or was there some other explanation? Certainly, a literal interpretation of Genesis would suggest only a four-day period for the whole of organic creation and the concept of fixed, immutable species was largely unquestioned in those days – although the eminent biologist, Lamark, had already proposed (incorrectly) that animals evolve through characteristics acquired during life and then pass these on to subsequent generations.

Darwin returned to England still puzzling about all he had seen. From his records it is clear that he was seriously considering the transmutation of species as a possibility as early as 1837. But if it occurred, what was the mechanism driving it?

In 1838 he began reading a remarkable book (for its time) by Malthus–the *Essay on the Principle of Population*. At a time of Enlightenment optimism, this was an uncharacteristically grim portrayal of an inevitable population explosion, curtailed only by famine and disease. It portrayed a struggle for existence for the human race. During this voyage, Darwin had seen population pressures at work in the animal kingdom. Could these be so great that only those most fitted would survive? It was clear that animals produce more offspring than survive to maturity. Moreover, these offspring often differ from each other and these differences are often passed down to later generations. Darwin also recognised that environments vary widely in terms of climate, food supplies and hostile organisms. Taking these three facts into account it is possible to see how, over time, a species might change to be able to meet the challenges and opportunities presented by the environment. Animals or plants inheriting unfavourable characteristics will, on average, produce fewer offspring that survive to maturity than those with characteristics favourable to the environment. This is 'natural selection' and 'the survival of the fittest'. Darwin went further. If this process were to go on for a very long time, there was the possibiliby of the change being sufficient for a new species to arise. And, given time, all living things on earth could have been descended from one or a few original forms.

Darwin first wrote up this theory in his notebook in 1842, and, more fully, in 1844. He then spent the next fifteen years developing and expanding it until eventually, on learning that Wallace had developed a similar theory, he finally published it in the book *On the Origin of Species by Means of Natural Selection* in 1859.

There was, of course, considerable opposition to the theory. One can identify four main lines of objection to it that have been raised over the years.

1) The theory cannot be proven

An objection that is often raised is that, because of the very nature of the theory, it cannot be subjected to the test of validity of a normal scientific theory–namely that it should be successfully tested many times independently by different scientists. It has also been claimed that there are circularities in some of the arguments–for example, survival of the fittest means little more than 'those who survive, survive'.

It has to be said that Darwin supported his thesis by thousands of facts that were available to him at the time. All that has been learnt in the subsequent 130 years supports the general line of the theory so that the basic science of evolution is securely established (even though some of its extensions remain controversial).

2) It is contrary to scripture

Clearly it is counter to the details of the creation stories in the early part of Genesis. But most Christians would accept that Genesis was written in a non-scientific age. Its purpose was not to give a twentieth-century account of science, but to convey the spiritual truth behind God's creation, the sinfulness and rebelliousness of man in spite of being made 'in the image of God', and God's justice, grace and mercy. Of necessity the language in which these truths are conveyed must be simple. Because creation is a special, unique and in the end unfathomable event, the language of poetry and myth is the appropriate vehicle to describe it. Most theologians would accept this view, pointing out that the rhythmic nature of the language is indicative of poetry, or at least enhanced prose, that Chapter 1 appears to be from a much later source than Chapter 2 and that the two accounts of the creation of humans in the first two chapters differ in detail.

However, many Christians, even today, would take exception to this view. In the United States in particular, there is a strong 'Creationist' movement which has battled for a balanced treatment in the schools so that where evolution is taught it should be accompanied (or even replaced) by a reading of Genesis as a *scientific* account of creation. I know Christians who hold this point of view, and have a high regard for their intellect and integrity. There are so many fundamental truths in the early chapters of Genesis that it is important to hold on to them. Moreover, it is clear that the authors of the New Testament were used to using the Genesis accounts to illustrate the new regime following the life, death and resurrection of Jesus. For example, in contrasting the new resurrection life in Jesus with the death introduced by the fall of Adam, Paul writes:

> For as by a man came death, by a man has come also the resurrection of the dead. For as in Adam all die, so also in Christ shall all be made alive (1 Cor 15:21-22).

Or again, in the Epistle to the Romans:

> Therefore as sin came into the world through one man and death through sin, and so death spread to all men because all men sinned—sin indeed was in the world before the law was given, but sin is not counted where there is no law. Yet death reigned from Adam to Moses, even over those whose sins were not like the transgression of Adam, who was a type of the one who was to come (Rom 5:12-14).

Clearly, if Paul had meant to convey the view that physical death originated with Adam, then there is a difficulty in reconciling this with the theory of evolution which requires the death of animals to have occurred over the hundreds of millions of years before man evolved and Adam existed.

However, most Christians would argue that it is the spiritual truths that are being conveyed which are all important here. We have all inherited a sinful nature in that there is a bias in us that makes us prefer to go our own way and ignore God. Humankind, as the animal fashioned in the image of God, has been given consciousness and the ability to choose. By choosing to ignore God we are cutting ourselves off from the source of life and opting for spiritual death–a deeper separation than is within the capacity of animals.

The resurrection life that Jesus brings gives us new spiritual life in this world as well as overcoming physical death. The whole body of Scripture testifies to these truths. Passages such as Ephesians 2:1–'And you he made alive when you were dead through the trespasses and sins' emphasises the present victory over spiritual death. They would also point to an overemphasis by the Creationists on God's early creative acts rather than on his continual creating and sustaining activity.

Paul writes of Jesus:

> He is the image of the invisible God, the first-born of all creation; for in him all things were created, in heaven and on earth, visible and invisible, whether thrones or dominions or principalities or authorities–all things were created through him and for him. He is before all things, and in him all things hold together (Col 1:15-20).

The creation relates to all things, not just to those created during the first week.

The author of the Epistle to the Hebrews gives a picture of the whole world upheld and sustained at this moment by the Creator:

> He reflects the glory of God and bears the very stamp of his nature, upholding the universe by his word of power (Heb 1:3).

They would therefore argue strongly that it is possible to be true to both science and to Scripture. Truly, evolution is God's solution to make intelligent life capable of responding to his love.

3) It eliminates the role of God

The Genesis stories, taken literally, suggest a four-day period for the whole of organic creation. In contrast, Darwin's theory suggests a continuous evolution over hundreds of millions of years, driven by the mechanism of natural selection. A superficial reaction from many people is that, whereas it would have been easy to recognise God's hand with a four-day creation period, it is not possible to do so with a continuous process like evolution. Why should this be? Are we all conditioned to think of God as an absentee landlord–having kick-started the universe into being and then left it–rather than the active Lord of creation?

Or are we falling into the trap of believing that, because we can see a possible mechanism, this necessarily rules out the hand of God in the process?

We shall have more to say about such issues in Chapter 6 when we consider God's provident action in the world. But for now we just claim that God's hand was on the evolutionary process, sifting the options, preparing the ground for his highest creation, humankind, to appear in the fullness of time.

4) It downgrades humankind

With a continuous line of evolution leading to *homo sapiens*, are we nothing but naked apes? This is another example of 'nothing buttery'. We shall see in the next chapter that, in spite of our biological continuity with the apes, we are much, *much* more!

Modern views on biology

The genetic code

Darwin's intellectual achievement in producing his theory was all the more remarkable because, at the time, little was known about the mechanism of heredity. Gregor Mendel was a contemporary of Darwin's. He was head of a monastery in Moravia and, by examining the ratios of the numbers of tall and short peas in successive generations, he was led to a numerical theory of heredity. His work was largely ignored during his lifetime, but by 1900 its significance in relation to the developing work on cells had been recognised. All living organisms are composed of cells—a single cell in some of the most primitive organisms to vast numbers in the larger, complex animals. The human body, for example, consists of 10^{13} cells of 210 different types from blood to bone, brain to muscle. Nearly all cells contain a nucleus which contains a number of threads, the chromosomes. In the human, there are twenty-three pairs of chromosomes in the nuclei of the body cells, but the egg and sperm cells have only twenty-three single chromosomes. When the egg is fertilised, the two sets of single chromosomes join to make the full set of twenty-three pairs in the new embryo. By the end of the 1930s, it had been firmly established that a large number of genes, determining the heredity characteristics, are strung out along the chromosomes like beads on a string.

The next step was to show that each chromosome is a single massive molecule of DNA. The structure of DNA was established by Crick and Watson in 1953 in the now famous double helix. DNA is an example of a macromolecule that can occur with carbon-based substances. Put at its simplest, it is composed of 'telegraph pole'-shaped units called nucleotides. There are four kinds of 'vertical' parts of the 'telegraph pole' called bases. The names of these bases, thymine, adenine, guanine and cytosine, do not matter for our purposes. We shall merely abbreviate them to 'T', 'A', 'G' and 'C'. The 'horizontal' part of the nucleotide is the same for all four types of nucleotide. The ends of the horizontal bar of the 'telegraph pole' are capable of bonding strongly with the ends of other bars to produce a stable long chain.

The exposed ends of the bases form relatively weak

bonds with bases of other nucleotides. So we end up with the ladder-like structure of DNA, the main structure being very strong but each rung containing two bases which can be readily split.

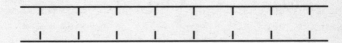

There is one more important chemical fact. The 'As' always bond with 'Ts', the 'Gs' with 'Cs'. It is now easy to see how DNA can replicate itself. If the molecule becomes unzipped, bare bases become exposed so that free nucleotides can fix themselves to them, 'Ts' always attaching themselves to 'As', 'Cs' to 'Gs', etc. This process is shown on the following page.

In human DNA there are some 10^9 base pairs in the twenty-three pairs of chromosomes. This means that there are 4×10^9 ways of arranging the bases in DNA and an incredible amount of information is stored in it. Before 1953, biologists regarded genetic information as stored in genes strung along chromosomes. Post 1953, it has become recognised that the molecular basis of this is the arrangement of base pairs along the DNA molecules that make up the chromosomes.

When we were conceived, half the information in our father's DNA fused with half the information in our mother's DNA resulting in our own special DNA contained within the nucleus of a single cell. This cell divided and subdivided many, many times in the formation of our body. But the nucleus of each of our cells contains the same replicated DNA molecules. With the vast information storage capacity of DNA, it is not surprising that, apart from identical twins, we are all unique in terms of our genes. On the spiritual level, we are each a separate, precious individual in God's eyes. This has important consequences for our views of genetic engineering, abortion and the dignity of the whole

1.

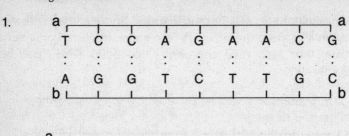

2.

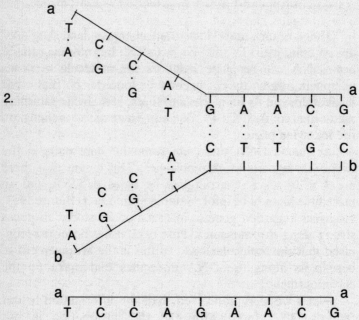

3.

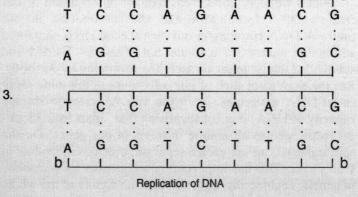

Replication of DNA

of living creation – but these issues are beyond the scope of this book.

We have seen, therefore, that our body is made up of a large number of cells of different types. Nearly all of these contain a nucleus consisting essentially of twenty-three pairs of DNA molecules which store the same information controlling the development of our body. How this is done is discussed in Appendix 6. But in essence it sends information via a substance known as RNA and therefore controls the synthesis of other macromolecules called proteins which are a major constituent of the cell. Some of the proteins called enzymes have the property that, because of their shape, they can speed up chemical reactions. One important function of some enzymes is to have the capacity to bind very stably to specific areas of the DNA, thereby preventing instructions to produce more protein from this gene to be repressed. It is through this type of control mechanism that a balance within the cell is maintained.

The protein factory

It is possible to envisage the cell as a complex protein factory. A recipe for the production of which type of protein and in which proportion is held within the DNA. Energy to keep the factory going is fed in from outside the cell through its semipermeable membrane wall. A chemical compound that plays a key role in this is ATP (adenosine triphosphate) which in animals is synthesised in a special part of the cell called the mitochondria. Enzymes carry the ATP to molecules that require an energy input to react so that the chemical energy which is stored in the ATP is released in just the right place in the cell to promote the necessary chemical reactions. On giving up its energy, the ATP becomes converted to ADP (adenosine diphosphate) – the ADP being reconverted to ATP in the mitochondria.

Information on the biochemical state of the cell is fed back to the nucleus and control exercised by triggering or repressing relevant parts of the DNA. When sufficient protein, nucleotides, cell wall material, etc, has been made, the cell is ready to divide. The DNA in the nucleus unzips as previously described, nucleotides attach themselves to the exposed bases and the DNA replicates itself, forming the heart of two nuclei of two cells.

The whole body

The above summary of cellular biology gives some feel of the intricacy and complexity of biological mechanisms as they are currently understood. This complexity is further increased when we consider the development of the whole body. We all start off as a fertilised egg–a single cell. At first sight we might expect to end up as a roughly spherical mass of identical cells as this first cell divides again and again. But, as we know, we are not like that. Even when the original cell divides, the resulting two cells are not identical. Within a few days, a hollow ball of cells has been formed and then a remarkable transformation takes place. The ball folds in on itself to leave an opening at one end. By two weeks, this process of gastrulation is complete, a primitive shape has been established and, signficantly, the cells destined to become skin, the cells that will give rise to the lining of the gut, those that will give rise to muscle, bone, blood vessels and those destined to give rise to the brain and nervous sytem have all been fixed. From this stage on, one cannot readily turn a cell that was destined to give rise to skin into one that was destined to give rise to the lining of the gut. This process is known as differentiation. As time goes on, cells become more and more differentiated and specialised into the 210 types we know in the human body. But the genes are identical in all cells in an individual.

How does this happen? What is it that tells an embryo to develop into a mouse rather than an elephant? How do the cells, having all started from a common ancestral cell, know to turn into liver, bone, muscle, brain? The essence of the answer seems to be that the basic instructions are programmed into the DNA. But these are not like a set of drawings. They are more related to the response to the chemical, electrical or mechanical potential gradients that exist within the developing body and will turn on specific parts of the DNA in the cells and cause each cell to develop in its special way. If you like, a cell senses where it is in terms of these potentials and reacts according to the programme which is universal to that body and encapsulated in the DNA. This process can be helped by physical contact between parts of the folded embryo. For example, in the human the embryonic eye first appears as a wine glass-shaped growth from the brain. The bowl becomes the retina and the stem, the optic nerve. As it grows it touches a uniform layer of cells that would form the skin of the head. On contact, these cells differentiate to form the lens of the eye.

There are other mechanisms. For example, when in Chapter 2 we were discussing chaos theory, it was pointed out that surprising patterns could emerge from the apparently chaotic behaviour of complex systems where there is an energy input. One such example is the Zhabotinskii reaction that is described in Appendix 2. Zhabotinskii mixed two specific chemicals in a container to result in a brown liquid. But this situation is not stable. He found that blue patterns would appear and spread, repeating themselves until the available energy was dissipated. Although the details will differ, it is possible that this principle may be applied in the growth of biological entities.

Whatever the mechanisms, it is clear that life as we know it is incredibly complex. Most biologists, whether Christians or not, view the wonder of the living creation with a sense of awe. How did it all happen?

The origin of life

The modern theory of the evolution of species by natural selection, taking account of our understanding of heredity, is known as neo-Darwinism. It argues that all life has evolved from a simple original form with DNA as the key information store. Evidence for this includes the fact that, today, the simplest and most complex forms of life use DNA as genetic material and, moreover, the basic code is always the same. For example GGC always means pick up a glycine molecule and AAG always means pick-up a lysine molecule whether in something as simple as a virus or as complex as man. Changes have occurred by two processes. When life had advanced to the stage of sexual reproduction, the existing genes from each parent are redistributed in the offspring. Secondly, some chemicals or radiation are capable of breaking the DNA chain which then rejoins in a new way, thereby changing the information sequence. Such a change is known as a mutation. The first process, sexual reproduction, is rather like shuffling a pack of cards. Mutations are like cutting the cards into small pieces and sticking them together to form entirely new cards. Over the longer term mutations are the more effective vehicle of change. Natural selection ensures that, out of this process, only the best hands of cards are retained. Given a few billion years, this process would lead to more and more complex life, adapted to survive in its particular environment. Ultimately it would lead to humankind.

But how did all this start? It is, of course, impossible to produce a definitive, proven, scientific theory, but there are interesting suggestions and speculations. The essence of the problem is that even the simplest living organism we see today is extremely complex and it is extremely unlikely that such an organism could have been formed by chance chemical reactions in the primitive earth. We have seen that two key ingredients of such an organism are chains of amino acids, the proteins, as well as a specific biological replicating material, DNA. Experiments have been undertaken to study how such an organism might have evolved. They are of two main types. The first category takes simple chemicals and sees how they might build up more complex compounds. It is believed that the early atmosphere of the earth contained water vapour, methane, hydrogen and ammonia but little or no free oxygen. Without oxygen present, these compounds could react to form organic molecules. Miller and Urey, in 1953, passed an electric current through such a gas and generated a wide range of organic compounds including amino acids (the building-blocks of protein). Other experiments have shown that starting with amino acids and energy from phosphates, in the presence of clays which have a large surface area to promote reactions, it is possible to build up protein-like chains. This is the chemical end.

The second main category of experiments start from simple biological compounds and show they are capable of growth and multiplication. Others start with RNA and show how it changes with environment in the laboratory. Although no complete mechanism for going from chemistry to life has been demonstrated, these experiments are illuminating. It may well be that no such demonstration will ever be made. That would hardly be surprising since the initiation of life must be a highly improbable event. There is no sign of fundamentally new forms of life appearing all over the earth. It is generally accepted that the first step–going from ordinary chemistry to the most elementary form of life is the most difficult one. Thereafter the properties of a living organism, that it is capable of multiplication, variation and heredity will permit higher forms to evolve by natural selection. Perhaps it has only happened once in the whole universe–we may possibly be one successful result of a vast number of attempts in all the planets. In this case, with so many attempts, the probability of success in an individual planet can be low. A range of speculations have been formulated, varying from an unlikely organic synthesis to inorganic material being used as a kind of initial template, and to at least two rather different theories by eminent scientists of life originating elsewhere in the universe and arriving on earth from space.

How proven is neo-Darwinism?

Karl Popper is the leading philosopher of science. He has pointed out that there is no absolute proof of a scientific theory. One can set tests for it. The theory is good only so long as it continues to pass all of these tests. From its very nature, the theories of evolution and natural selection are impossible to repeat in their entirety because of the unique nature of the development of life. How far is neo-Darwinism well established, how far is it speculative and how far is it merely a gleam in the eye?

First of all, there is strong scientific evidence to support the theory where the evidence is available – this relates primarily to a relatively recent time frame. For example, it is possible to trace the formation of new species in the case of the herring gull. In North America, they are somewhat different to those in the UK, and these differences progressively increase as one proceeds westward through Siberia, Northern Europe and back to the UK where we find a different bird has arisen, the lesser black-backed gull. The differences have accumulated to the extent that a completely new species of bird has appeared.

On a grander scale, the evidence rests on the observed fossil sequences. Fossils only form in suitable environments so, necessarily, the record is incomplete. But the amount of evidence is vast.

Over 7,000 different organisms, plants and animals have now been documented and there is clear evidence that the more primitive life first occurred in the older rocks, the more advanced in the newer ones. Dating the rocks using radioactive techniques one finds the sequence of first appearances.

Million years ago

First life	3,400
Metazoans	570
Shelly marine invertebrates	500
Jawless fish	510
Land plant spores	440
Jawed fish	420
Amphibians	360
Reptiles	320
Coniferous plants	300
Mammals	210
Birds	150
Flowering plants	120
Early hominids	7
Tool-using ancestral man	2
Homo Sapiens	0.1

If we assume that God did in the past what we observe him doing today, then it is reasonable to suppose this sequence can be accounted for by neo-Darwinism. It is the best, and indeed the only credible, scientific theory that we have. There is some evidence to suggest that evolution may have proceeded in a series of stops and starts but this does not affect the overall argument.

Then there are issues like how specific features such as eyes have developed. Here it is possible to suggest a series of steps (each one of which had survival benefits) that could have led to the development of the eye that we enjoy today. But there is a degree of speculation rather than firm proof in this.

Finally, there is the issue of how life began in the first place. As we have seen, there are suggestions of how it might have happened but these are all highly speculative. It is hard to see how we will ever be able to get hard information to raise such speculation to the level of a well-established scientific theory.

Neo-Darwinism remains an elegant theory but it may not be the whole story. Protagonists such as Dawkins argue passionately that it is. In his book *The Blind Watchmaker* he writes:

> At first sight there is an important distinction to be made between what might be called 'instantaneous creation' and 'guided evolution'. Modern theologians of any sophistication have given up believing in instantaneous creation. The evidence for some sort of evolution has become too overwhelming. But many theologians who call themselves evolutionists, for instance the Bishop of Birmingham, smuggle God in by the back door: they allow him some sort of supervisory role over the course that evolution has taken, either influencing key moments in evolutionary history (especially, of course, *human* evolutionary history), or even meddling more comprehensively in the day-to-day events that add up to evolutionary change.
>
> We cannot disprove beliefs like these, especially if it is assumed that God took care that his interventions always closely mimicked what would be expected from evolution by natural selection. All that we can say about such beliefs is, firstly, that they are superfluous and, secondly, that they *assume* the existence of the main things we want to *explain*, namely, organised complexity. The one thing that makes evolution such a neat theory is that it explains how organised complexity can arise out of a primeval simplicity.
>
> If we want to postulate a deity capable of engineering all the organised complexity in the world, either instantaneously or by guiding evolution, that deity must already have been vastly complex in the first place. The creationist, whether a naive Bible-thumper or an educated bishop, simply *postulates* an already existing being of prodigious intelligence and complexity. If we are going to allow ourselves the luxury of postulating organised complexity without offering an explanation, we might as well make a job of it and simply postulate the existence of life as we know it!

What does the Christian say to such an attack? In the first place, neo-Darwinism does not explain organised complexity as Dawkins claims. It relies on the laws of nature being such that fruitful complexity can develop, and we have seen in Chapters 2 and 3 that this in turn requires that we should live in a very special universe that is finely tuned. Dawkins gives no explanation of how this came about. Moreover, if we make the reasonable assumption that this is the workings of our God, then is it not also reasonable to suppose that the same God would be at work through the long evolutionary process? It is only by putting a box around biology and ignoring everything else that Dawkins can discuss God in such a cavalier fashion.

Secondly, if we humans are merely the product of random mutations sifted by the processes of natural selection, how can we have the capacity to know this is the nature of reality? Are our minds, which Dawkins assumes are merely one of the products of this process, capable of standing outside the process and coming to a valid judgement about it? This is a difficult philosophical question that Dawkins does not appear to address. It brings us to the key issue of the nature of humankind–the subject of our next chapter.

5

In the Image of God

'So God created man in his own image, in the image of God he created him; male and female he created them'
(Genesis 1:27).

Introduction

ONE INTERESTING RESULT from the comparative studies of the DNA in humans and apes is that it is possible to determine which species of ape we are most closely related to and obtain some idea of when the species diverged. It seems likely that we are most closely related to chimpanzees and that the two lines diverged between seven million and five million years ago. The animals along the line leading to humans are known as hominids. The early hominids left the trees and lived in open country in Africa. They were ape-like with small brains and known as *australopithecines*. Significantly, they were bipedal–walking erect.

By two million years ago a new hominid had arrived– *homo habilis*–with a significantly larger brain than *australopithecines*. At this time we have the first evidence of the use of primitive tools.

Homo habilis was replaced over one and a half million years ago by a slightly larger-brained hominid–*homo erectus*. This was the first hominid species to migrate outside of Africa. Their ability to make stone tools and implements shows the extent of the development of manipulative and mental skills that had developed since the early hominids.

Then around three hundred thousand years ago, hominids similar to, but with somewhat larger brains than *homo erectus* began to appear. These have been called *'archaic homo sapiens'*, the best known coming from Europe and West Asia–the Neanderthals. Anatomically, they were stocky and powerfully built. They were capable of making high-quality stone tools–many of the later ones being hafted as spears and knives. But it seems that they were scavengers and hunters of, at most, small game and unlikely to be capable of planned logistic big-game hunting in the way that modern humans are. All the evidence points to their communication, presumably speech, being much less effective than that of their successors–modern *homo sapiens*.

A significant change occurred over the period between one hundred and thirty and thirty thousand years ago, ending with the emergence of humans like ourselves, *homo sapiens*. Genetic studies suggest that we could have appeared first in one geographically restricted area–probably Africa–and spread from there.

In a purely biological sense there is a continuous chain between *homo sapiens* and other animals–we still share 98% of our genes with chimpanzees. But, even in a purely scientific sense, the emergence of modern man signalled a major qualitative change.

In the first place there is self-consciousness which is unique to humans (there is some evidence for a knowledge of self in the chimpanzee, but this is of a very primitive kind). The higher animals are conscious but not self-conscious.

There is general agreement that the ability to communicate by language was a significant factor in the transition. There was the use of sophisticated tools, co-ordinated and co-operative hunting, a growth in population size and density, and the use of cave painting and sculpture. The potential was there–a potential that was later realised as humans changed from hunter gatherers to farmers to living in large communities.

Christians would claim that this potential included an emerging capacity to be in the 'image of God'. What does this imply? This is, of course, a major theological issue but it is possible to identify four main attributes:

1) An intellectual capacity that includes the ability to understand.

2) A capacity for responsible choice. This includes free-will and the ability to plan and thereby shape the future.

3) A moral capacity, including the ability to love others– a feature shown supremely in the life of Jesus.

4) A capacity to know God–to understand his purposes, to follow his leading, to be strengthened by his Spirit, to be with him throughout eternity.

How far is this supported by the insights afforded by modern science? We shall be reviewing the evidence in the rest of this chapter.

We start by reviewing briefly what we know of the mechanism of the brain. Then we summarise various philosophical views on the distinction between a brain and a mind. Next we ask the question whether there is any difference in principle between a human brain and a computer. We go on to examine views on the origin of behaviour and language. Finally we review the implications of this for our Christian faith.

The human brain

If there is going to be a significant qualitative difference between humankind and the animals, there is general scientific agreement that it is likely to be associated with the brain.

Brains are made of cells called neurones. There may be only a few hundred cells in some simple invertebrates to as many as a hundred billion (10^{11}) in humans. From the cell body of a neurone emerge relatively short dendrites which pick up signals from other neurones. There is also a single axon, which can be a metre long, which carries signals away from the neurone. The signals are transmitted electro-chemically in the form of pulses. A neurone will emit pulses along its axon at a firing rate of up to several hundred per second, the rate depending on the strength of the stimuli reaching its dendrites from other neurones.

A significant feature of the brain is the large number of connections between neurones via junctions known as synapses. The dendrites of sensory neurones make contact with a sensory cell such as a light-sensitive cell in the retina. The sensory neurones, therefore, provide an input to the brain. The output is via motor neurones, axons of which make contact with a muscle fibre. But most neurones in the brain make contact only with other neurones. It is the development of these connections that is, at least in part, associated with the process of learning.

The part of the brain that is more developed in humans than other animals is the cerebrum covering most of the outer parts of the brain. It consists of a comparatively thin outer layer of grey matter, the cerebral cortex, overlying larger inner regions of white matter. The grey matter is where various kinds of computational task appear to be performed while the function of the white matter appears to be to carry signals from one part of the brain to another. The cerebral cortex is divided into two hemispheres which com-municate with each other through a part of the brain known as the corpus callosum. Specific areas of the cerebral cortex are associated with the receipt of stimuli from various parts of the body. Strangely, it is the left hemisphere that often receives information from the right-hand side of the body and the right hemisphere from the left side. Information from the right-hand field of vision of both eyes goes to the left hemisphere. The two speech centres are normally con-centrated in the left hemisphere, one being concerned with comprehension, the other being concerned with the formula-tion of speech.

Although the cerebrum is particularly well developed in

humans, there are other important parts of the brain. They are not crucial for our discussion, but for completeness we mention some of the more significant ones.

The cerebellum:	concerned with the precise co-ordination and control of the body. We do many things automatically and the cerebellum controls this. When we learn a new skill we need to think what we are doing (cerebrum in control) but when we have mastered it the cerebellum takes over and we may not perform as well if we stop and think what we are doing.
The hippocampus:	plays a vital role in the laying down of long-term memories.
The hypothalamus:	the seat of emotion.
The thalamus:	processes many of the inputs from the external world and transmits them on to the cerebral cortex.
The reticular formation:	responsible for the general state of alertness or awareness in the brain.

With this general introduction, let us proceed to examine how the brain perceives the external world. Most is known about the processing of visual information. Light falls on the retina of the eye and some processing of the signals takes place within the retina itself before proceeding to the part of the cerebrum known as the visual cortex. It seems that the cortex is extracting specific features of the image on the retina–for example, a sloping line at a certain angle–before analysing the information as a whole. It requires an enormous jump to move from a collection of elementary information extracted from the two-dimensional image on the retina to discerning that one is looking at a three-dimensional object that one can recognise (that is a lady whom I recognise to be my wife!). The brain has a truly remarkable computing power to be able to undertake this task. It would seem to entail storing the essential features of objects in the brain and comparing incoming signals with the stored information.

Because of the way the brain treats the data, it is possible to obtain ambiguous or false perceptions of a visual image. The illustrations show some well-known examples of this phenomenon. Those who take the view 'unless I can see I won't believe' may not be on so secure a ground as they care to imagine!

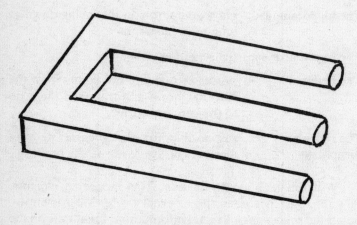

A nonsense picture

The conscious mind

With this simple account of the workings of the brain we go on to consider one of the features that distinguishes humankind, our developed self-consciousness. Here there is something that seems to stand outside the mechanistic electro-chemical computer we call the brain and yet is uniquely

Are the lines bent?

associated with it. If I am truly human I am aware of myself, I can perceive beauty, I have a moral sense, I have the ability to think creatively about the future, I can will my body to undertake tasks that I have chosen, I can love others, I can respond to God's love to me. Although most people would accept we have a conscious mind, it is not easy to define it precisely or understand its relationship to the brain.

The relationship between the brain and mind is a topic that has exercised philosophers over the centuries. Although there have been many different approaches, none has been entirely satisfactory and won general acceptance. One can identify three major strands.

The first is Dualism, often associated with the French philosopher Rene Descartes (1596-1650). He considered the mind and body to be two quite separate but interacting entities or 'substances'. His approach was attacked, in 1949, by the Oxford philosopher Gilbert Ryle who coined the 'deliberately abusive' slogan–'Dualism is the theory of the Ghost in the Machine.' He claimed that Descartes had made a category error–an analogy was that of a visitor to Oxford University who, having visited all the colleges, asked if he could now see the university. The latter is the sum of the former, and so, according to Ryle, is the relationship between mind and brain. But one suspects that a major reason for the loss of popularity of dualism rests with its counter-intuitive nature. Our culture is still one in which one needs to see to believe. It is ironic that with that most precise scientific discipline, physics, we in the twentieth century have come to appreciate that the fundamental laws governing the behaviour of matter are dualistic in character (see Appendix 3). Indeed, Sir John Eccles, a Nobel prize-winner for Medicine, has suggested that the mind-brain interface could be linked by the displacement of a minute particle which is subjected to the uncertainties of quantum physics by the Heisenberg Uncertainty principle.

It is important to note that the Bible does not adopt a dualistic view of humankind. Each individual is a psychosomatic unity. When the Bible refers to soul (*nephesh* in the Old Testament, *psyche* in the New), it is referring to the whole person, not a part.

The second is Behaviourism. The behaviourists choose to analyse a person's actions not from a mental activity such as an act of will or belief, but as a response to stimuli. If it is raining outside, I will respond by getting my umbrella. The behaviourists would argue that there is nothing more to it than this.

The third is Materialism. The materialist considers the brain to be normal matter. What we regard as conscious thoughts are merely brain states. But there is strong evidence from radio-tracer experiments that there is widespread neutral activity when a person is simply thinking in the absence of any external stimulation. The materialist would find this hard to explain.

There is one further piece of evidence that should be mentioned. Some patients with severe epilepsy have had an operation in which the corpus callosum, the tract linking the two cerebral hemispheres, was severed. They were left with, in effect, two separate unconnected brains, each half being capable of performing its original task. The question arises as to where self-consciousness resides. The most careful investigations indicate that although not totally independent, both halves of the brain are separately conscious but the major attributes of self-consciousness are concentrated in the left hemisphere. If there had been two separate fully self-conscious minds in the same individual, this would have raised difficult issues.

Merely a computer?

We live in an age when medical technology has advanced to the stage where successful organ transplants take place. If I have someone else's kidney or heart, I still feel that the body I have is really me. If asked which part of the body is most associated with the 'real me', most would accept it is the brain. But physiologically the brain is essentially a powerful computer. So the question arises as to whether it is feasible to build a computer to resemble the brain. And if we could would it be conscious and have a mind?

The brain is, of course, a very special computer. It does not rely on the silicon chip but is electrochemical in its operation. It is also strongly connected internally. But the British mathematician, Alan Turing, has showed that, in principle, all computers can be simulated on a theoretical computer known as a Universal Turing Machine. Some computers work more rapidly than others, some operate on valves, some on micro chips, some on punched cards, but, at the end of the day, they are all capable of performing the same type of logical operations.

We are well aware that computers are capable of undertaking calculations much more rapidly than the human brain and computer technology is revolutionising our offices and commercial life. They may be good for storing information, routine calculations and so forth, but what about the human ability to appreciate beauty, to experience feelings, to love? Would a computer have a mind? Would it be self-conscious?

These are questions that have received some considerable attention in the last couple of decades or so. A whole new field, cognitive science, has emerged as an amalgam of psychology, philosophy, mathematics, linguistics, computer science, anthropology and neurophysiology. The issues raised are difficult ones and there is often strong controversy between those working in the field. Christians should be cautious of accepting strongly-worded and often plausible claims to have developed a breakthrough in understanding in this area. I will illustrate some of the controversy by two examples.

The first example is, if it were possible to build a computer that really thinks for itself (as opposed to merely doing what the programmer tells it) how would you know? Turing suggested the following test. Suppose you were in a closed room but could send messages and receive messages from outside. You do not know whether you are going to communicate with a person or with a computer. You can ask any question you like and the person or computer replies. If, from the nature of the replies you receive, you cannot distinguish whether it is a computer or an individual who is replying, and if it transpires that, in fact, you were communicating with a computer, then it would be hard to claim that the computer does not think. Naturally the most testing questions would be those where understanding and judgement is involved. A computer would perform better than a human at calculations or factual questions of the 'mastermind' type. One might think that such a test is simple and straightforward.

But the principle of this approach has been challenged by the American philosopher, John Searle. He postulated that the questions passed to the computer should be written in Chinese. A person who only speaks English should receive the questions and, according to a book of instructions written in English (the computer programme) should manipulate the strange Chinese symbols and, following instructions, post out a sequence of Chinese characters as the reply. The point is that the operator does not understand the question (because it is in Chinese) and yet he is capable of giving a reply which might pass the Turing test. Therefore he concludes that passing a Turing test does not imply that a computer is thinking in a conscious way.

Searle's argument raised a great furore among some of the members of the artificial intelligence community. Hofstadter wrote it was 'one of the wrongest, most infuriating articles I have ever read in my life' and regarded it as 'a religious diatribe against artificial intelligence'.

The second example is that of a remarkable piece of mathematics known as Gödel's theorem. In the early part of this century, mathematicians were developing an abstract form of mathematics known as formal systems. The idea was to start with some assumptions (axioms) which were fairly general but precisely defined and, using a set of logical rules, to identify all the possible theorems (the results of applying the logic) that could be derived from the initial assumptions. Then in 1931 the young Austrian mathematician, Kurt Gödel, shook the mathematical and philosophical world by proving that, in such a system, there are truths that can be known but not proved. In other words, even in as rigorous a subject as mathematics, there is a limit to the amount of truth that logic can extract from the system. Now a computer with its programming is just a formal system. So a computer can only get at a certain amount of truth.

This point has been powerfully argued by Professor Roger Penrose in his best seller *The Emperor's New Mind*. He shows that any computer would be incapable of solving certain mathematical problems that humans can solve. We might add that, if we can perceive mathematical truth that is beyond the reach of a computer, then it is unlikely that computers will ever attain the aesthetic, emotional and spiritual capacity that we all possess.

Windows into the mind

So far in this discussion we have examined the evolution of man and looked briefly at the neurophysiology of the brain. Then we summarised the differing philosophical views on the relationship between mind and matter before examining the inconclusive evidence on whether the brain is merely a super computer. Now we turn to three different scientific views as to why we behave as we do. It is important to stress at the outset that the three views are only partial ones. They may each be the scientific truth and nothing but the truth but they are certainly not the whole truth. They are best regarded as three glimpses through separate windows into the workings of the mind and three different pictures emerge. There have been acrimonious debates between the scientists who have sought to claim that their particular view was the only way to approach the subject.

Christians could legitimately claim that they too have a special view into why we behave as we do. That adds to, but need not conflict with, the three scientific ones described below.

Sociobiology

The first glimpse into human behaviour is through the window of sociobiology. Our behaviour is determined by the genes we inherit. Moreover, if this behaviour is such that our genes multiply and survive in succeeding generations more rapidly than other people's genes with a differing behaviour pattern, then our behaviour pattern will become the norm. So, according to this view, our behaviour is the result of Darwinian selection.

Sociobiology is very good at explaining the innate behaviour of animals. Young birds in their nest are programmed to open their beaks when their parent appears with food. Those with genes that did not cause them to do this would starve and this particular gene would disappear with them.

But the big problem that sociobiology faces is the problem of good. By and large people wish to live in a society that is based on justice and a concern for others rather than the survival of the fittest. Although sociobiology can account for altruistic behaviour in some animals (why bees will sting an enemy and die to protect the queen for example – the answer lies in the fact that they share the same genes) and the concern of parents for their children, it is very hard to account for our desire for goodness from sociobiology alone. Richard Dawkins, a prominent sociobiologist, has written:

> My own feeling is that a human society based simply on the gene's law of universal ruthless selfishness would be a very nasty society in which to live.... Be warned that if you wish, as I do, to build a society in which individuals cooperate generously and unselfishly towards a common good, you can expect little help from biological nature. Let us try to TEACH generosity and altruism, because we are born selfish.

The Christian would accept this as a correct diagnosis of the human condition. But he would point out that teaching generosity and altruism, although good, is insufficient. Our very selfish nature that Dawkins recognises will pull us down. St Paul understood this when he wrote:

> For I know that nothing good dwells within me, that is, in my flesh. I can will what is right, but I cannot do it. For I do not do the good I want, but the evil I do not want is what I do.... Wretched man that I am! Who will deliver me from this body of death? Thanks be to God through Jesus Christ our Lord! (Rom 7:18-20, 24-25).

The heart of the Christian gospel is that, although we cannot pull ourselves up, Jesus can transform us.

Behaviourism

An entirely different approach is that of the behaviourists. They deliberately do not attempt to understand what is going on inside the mind, but only relate behaviour to external stimuli. Perhaps the best-known example is the work of the Russian, Pavlov, who suggested the concept of a 'conditioned reflex'. An unconditioned reflex is one which appears without previous experience–if food is placed in a dog's mouth it salivates. Pavlov found that if some other stimulus, such as the ringing of a bell, always just proceeded the giving of food, then the dog would salivate in response to the ringing of the bell alone. Such a response is a conditioned reflex.

Behaviourism became fashionable in the 1920s as a result of J.B. Watson and his book *Behaviourism*. He wrote:

> Give me a dozen healthy infants, well formed, and my own specified world to bring them up in and I will guarantee to take any one at random and train him to become any kind of specialist I might select–doctor, lawyer, artist, merchant-chief and, Yes: even beggar-man and thief,

regardless of his talents, penchants, tendencies, abilities, vocations and race of his ancestors.

In essence, behaviour will be determined by rewarding desirable actions and punishing undesirable ones. In this way one builds up the desired conditioned reflexes. Clearly one can influence behaviour in this way but it is hard to account for originality or purpose. Did the genius of a Shakespeare or a Mozart arise as a result of conditioned reflexes? Surely not!

It was because of this type of difficulty that B.F. Skinner was led to the concept of 'operant conditioning'. In 'operant conditioning' the sequence 'first stimulus, then response' is reversed. Desired behaviour is achieved and maintained by positive reinforcement (rewards) after a correct action rather than before it. Starting from a random tendency to do X, a behaviour pattern is built up in a series of steps. The reward can be given by a person or by the environment. The analogy between operant conditioning and Darwinian evolution by natural selection is obvious.

For Watson and Skinner there is no room for the mentalist concept that behaviour is caused by feelings, ideas, wishes or intentions. They considered such concepts to be unhelpful and non-explanatory. But it is important to note that they did not disprove them. By only considering external stimuli and responses they ensured they could not discover the importance of the mind. Later behaviourists have gone some way to remedy this omission by the development of arousal theory.

Operant conditioning is no longer as prominent as it was forty years ago. One of the reasons for this is that attempts to use it to explain language acquisition were far less satisfactory than the alternative approach suggested by Noam Chomsky. The ability to communicate ideas by language is a unique human attribute. It is hard to exaggerate its significance in the development of thought and the growth of civilisation. But children seem to be able to master a language without significant instruction. This process takes place most rapidly between the ages of two and three. Moreover, it seems that a normal human child is able to master any human language to which it is exposed in infancy. This led Chomsky to suggest that the human brain contains a genetically programmed 'language organ' enabling children to master their mother tongue with virtually no training or effort. Moreover, this 'language organ' is capable of providing the foundation for all known languages. Only those languages that fall within the scope of this innate capability are suitable for human communication. In effect, in each infant's brain there is an in-built framework which contains the foundation of all languages–the specific language to which he is exposed can then be rapidly acquired as a filling in of the framework. Operant conditioning is far too slow a process to account for language acquisition.

Psychodynamics

A further insight into human personality has been obtained by psychiatrists, treating patients with psychotic or neurotic disorders. The best known of these is, of course, Sigmund Freud. Freud's early work occurred at a time when classical physics was at its peak (with its emphasis on energy) and Darwin's theory of evolution was widely accepted. He sought to develop a theory of the personality also based on forces and energy–these acting at a level deeper than that of the conscious mind. Freud considered that the total personality is composed of three main systems: the id, the ego and the superego. The id is the primary source of psychic energy and seeks to gratify basic instincts. Freud laid great emphasis on satisfying the sexual instinct which he considered to be a potent force, even in infancy.

The second system is the ego–the rational, controlled part of our personality which seeks to keep the instinctive behaviour of the id in check.

The third system is the superego which is the moral or judicial part of our personality. It is, if you like, our ideals and conscience. The superego seeks to influence the rational ego which, in turn, seeks to control the instinctive impulses of the id. It is the history of the interaction between these systems over a person's life that Freud considered to lead to an individual's personality, including his neuroses.

There is, undoubtedly, a great deal of truth in this view which forms the basis of much of the modern method of analysis of interpersonal relationships as well as psycho-analysis. But, taken to extremes, it can give rise to the false view that our behaviour is merely the product of the sum of our experiences over life – that we are not in the end account-able for our actions. A Christian of any sensitivity would recognise that a person's personality will be influenced by his environment but, at the end of the day, will insist that the root cause is that we are all sinners and in need of the grace of God.

For Freud, seeking to develop a scientific model of the personality, religion was merely a delusion. This was far from the view of another well-known pioneer of psychology, Carl Gustav Jung. For a time Jung worked closely with Freud, but in 1913 the two diverged. It is to Jung that we owe the concept of the collective unconscious. Beneath the unconscious layer which, in the Freudian view, is formed from accumulated individual personal experience, is a deeper level of the unconscious, the contents of which are more or less common to all humankind. Jung called the content of the collective unconscious, archetypes. He consid-ered that they formed the foundation of our personality and behaviour. For example, the archetypes animus and anima relate respectively to a woman's image of a man and a man's image of a woman. They are concerned with the male and female personality as a whole, not merely sexual aspects. Another archetype is the 'wise old man' representing supe-rior insight. Jung's descriptions were often somewhat vague and it is not obvious how this differs from a father-figure or even God. Jung derived his archetypes from his experience of psychoanalysis (including his own) and from the study of myths and comparative religion. Although Jung was not a conventional Christian believer, a summary of his views would be that, within each of us, there is an innate capacity and need for belief in 'God'. The Christian would assert that this need is only fully met by 'the God and Father of our Lord Jesus Christ'.

We all have the equipment within us to be able to believe in God in just the same way that Chomsky was able to show we are born with a capacity to learn any language.

St Paul was completely in tune with this when he spoke to the Athenian philosophers (Acts 17). Judging from their idols, they had a great hunger for religion which St Paul claimed could only be satisfied by the one true God.

An extra dimension

So far this chapter has been concerned with reviewing some of the diverse scientific and philosophical approaches to a study of human nature. From this wide-ranging and, in places, contradictory account one could be excused for believing that this is not the whole story. There must be more to it.

It will be recalled that we started by considering the evolution of man and found a biological continuity with other primates. It raises the question of whether we are just animals or whether there is something very special about us that makes us fundamentally different from them.

Then we looked at the neurophysiology of the human brain and found it to be capable of the most remarkable analysis. But most people would accept that our total personality is determined by much more than just a collection of interconnected brain cells and we are very much more than an elaborate computer.

We looked at some of the attempts that have been made over the centuries to explore the relationship between our conscious minds and our bodies and found there was no consensus of view between philosophers.

Then we considered sociobiology, behaviourism and

psychodynamics and found that each gave a useful but only partial insight into human nature.

All these investigations have one thing in common – they all start with man as he is and attempt to analyse one aspect of his nature. They provide a bottom-up approach to the question of what is human nature. Can we get any further by the top-down approach offered by Christian theology? We should not, of course, expect to discover detailed twentieth-century science in such an overview. Rather we should be discovering purpose and meaning, thereby providing a framework into which the scientific insights described earlier in this chapter can be fitted.

We may look again at the four attributes of being in the 'image of God' mentioned earlier in the chapter.

1) An intellectual capacity that includes the ability to understand.

We have seen the structure of the powerful human brain which is capable of a perception of even mathematical truth which is fundamentally inaccessible to any computer. Humans have an ability to be creative and to communicate and to conceptualise which is well beyond the capacity of any other animal. We can seek to express meaning through the arts and are able to look beyond our own direct experiences and wonder about the meaning of life and death. Biologically we may be animals but it is clear, even at this stage, that we are very special ones.

2) A capacity for responsible choice. This includes freewill and the ability to plan and thereby shape the future.

The Christian would insist that we are responsible beings. There may be much truth in behaviourism, sociobiology and Freudian psychology, but they all fall down in failing to recognise that, at the end of the day, each one of us is accountable (indeed, accountable to God) for his own actions. Of course we cannot be blamed for those parts of our behaviour resulting from being trained badly (behaviourism), having inherited a weak nature (sociobiology) or having had terrible early experiences (Freud). But, given the stack of cards we have been dealt, we then have both the freedom and duty to play them well.

3) A moral capacity–including the ability to love others–a feature shown supremely in the life of Jesus.

Deep down within us we all recognise the virtues of justice, honesty and straight dealing, whether or not we practise them. One might be able to argue that these were essential for civilisation to develop and so the successful communities were the ones who took steps to encourage such qualities. Or, alternatively, one could contend that humans were intended by God to have a moral capacity. But I would suggest that the Christ-like conception of love is not simply explained away. There are no selection pressures favouring the emergence of loving creatures–those that seek the well-being of others rather than themselves. Jesus said: 'A new commandment I give to you, that you love one another; even as I have loved you, that you love also one another' (Jn 13:34). Notice the word 'new'. Love does not come naturally. 'Greater love has no man than this, that a man lay down his life for his friends' (Jn 15:13). At the highest level, we humans are to live lives which are directly opposed to what one would expect if we are merely special animals developing under rules of the survival of the fittest.

4) A capacity to know God–to understand his purposes, to be strengthened by his Spirit, to be with him throughout eternity.

Central to the whole of the preceding discussion is the Christian view that human beings are God's creatures. All our remarkable abilities become meaningful once we recognise that our prime purpose in life is to serve and worship the living God.

And God has equipped us for this task. Recognising this puts the scientific approach into perspective. Of course we have been given a wonderful brain. Of course we have freewill–which includes the freedom to ignore and reject God as well as to put him at the centre of our lives. Of course we will tend to grow and flourish when nurtured within loving relationships–and most especially when we understand something of God's love for us shown in the life and death of Jesus. Of course we expect to find, as indeed we do find, in humans through the ages an urge to worship something beyond themselves–the God-shaped hole that must be filled.

One cannot give an absolute proof that humans are made in the image of God, but looking at what we know of humankind from our scientific studies makes sense when viewed from this perspective.

6

God in Action

'*Behold, the Lord's hand is not shortened, that it cannot save, or his ear dull that it cannot hear; but your iniquities have made a separation between you and your God, and your sins have hid his face from you so that he does not hear*' (Is 59:1-2).

Can miracles happen?

ONE OF THE GREAT THEMES of the Christian religion is that God is active in his world. At the beginning of time he speaks and the universe is created. He calls Abraham and founds a people of God. He leads this nation out of captivity from Egypt, parting the Red Sea so they can cross it. He is the God of battles, leading to the settlement in the Promised Land. He demonstrates his power time and time again, but eventually allows the rebellious Jewish nation to be carried into captivity in Babylon. It is claimed that God even acts through the pagan Cyrus to defeat the Babylonian Empire and thereby enable a faithful remnant of the Jewish people to return to Jerusalem.

God acts supremely by sending his Son into the world. After a three year ministry, Jesus is crucified and God raises him from the dead.

The Holy Spirit – God in action – comes to the bewildered followers of Jesus just as he had promised, turning them into a rapidly expanding church which turns the Roman world upside down. Over the centuries and even until today Christians can testify to the transforming power of God in their own lives.

Such is God's activity at its most dramatic. God also acts through the normal daily lives of his faithful followers. He is there in the house, the workplace, the sports-field, the supermarket. Belief in a powerful, concerned, loving God, upholding his world, is at the core of our Christian faith.

However, many people today, brought up in our Western post-Enlightenment culture, find it hard to accept that God can act in this way. This issue is seen in its most acute form with miracles. And yet the evidence for, say, the physical resurrection of Jesus is extremely compelling. What happened to his body if there was no resurrection? If the Jews or Romans had stolen it, they would surely have said so when faced with the uncomfortable impact of early Christianity. If the disciples were fabricating a story, surely they would not have been prepared to die for what they knew to be a lie. And if we accept the key miracle of the Resurrection, why should we not be prepared to accept the accounts of the other miracles?

But it is often said that miracles are unscientific. In one sense this is correct, but in another sense quite illogical. Science only deals with the regular. The scientist observes the behaviour of things (living or inanimate) in the universe and deduces patterns of behaviour. Experiments, which can be repeated by different scientists in different parts of the world, are the foundation stone of our scientific knowledge. So science deals only with the regular and the repeatable.

This is in stark contrast with a miracle which, by definition, is an unusual event with spiritual significance. So miracles do not fall within the ambit of science. Science can have nothing to say about their validity. It is only in this way that miracles are unscientific.

This is a simple answer to a simple question. It is treated in this way because it is the question that is so often raised in discussion. But there are much deeper issues concerning God's action in the world. These include:

- The problem of suffering: why doesn't God intervene more often than apparently he does?

- Does God's action mean that scientific laws are necessarily broken?

- Is it possible for God to act without it being apparent to everyone?

The problem of suffering

We live in a world where there is manifest cruelty and injustice as well as suffering due to natural disasters and diseases. The basic question is why a loving, almighty God should permit this to happen. If one is a Creationist, the answer is that such suffering only arose as a result of Adam's sin at the fall. So floods, earthquakes, etc, only arose after man had first sinned.

Those Christians who accept that the world is billions of years old and that humankind has evolved from primitive animals at the end of a long period of evolution are faced with a more complex problem. Natural death in the animal kingdom must have occurred for hundreds of millions of years before man first sinned. Clearly, in this view, physical death could not have arisen as a result of man's sin. But spiritual death—a separation from a close relationship with God—could. So the suffering caused by human greed, selfishness and lack of love can quite legitimately be put down to human sinfulness.

But there are also 'natural' disasters—earthquakes, tidal waves, etc. Moreover, if we accept the concept of evolution, we are faced with a process lasting hundreds of millions of years where every advance has been won at the expense of those creatures who failed to make the grade. The strategy of evolution through natural selection is a very selfish and profligate one. There have been notable 'dead ends' like the dinosaurs. How does all this stack with the Christian view of an all-powerful and all-loving God who is active in his world? Could he not have created a better world?

We are all aware of the desperate famines that occur in various parts of the world. Why does the God of miracles permit these? Where was God when millions of Jews were being exterminated in the Nazi concentration camps? Why do some people seem to progress smoothly through life while others face major problems from early childhood? We are all aware of the issues. And there is no complete intellectual answer to them.

Jesus made it clear (Lk 13:1-5) that such disasters do not fall predominantly on the worst sinners. He illustrated this point by discussing the case of some Galileans who were killed by Pilate while they were offering sacrifices to God, and also that of the eighteen people who were killed when a tower at Siloam collapsed.

In John 9:1-7 we read of a similar question being asked about a man who was born blind. Jesus explained that this affliction was not due to the sin of either the man himself or his parents. Jesus' action was to heal him. He had a better answer than a merely intellectual one.

In the Gospels we read of Jesus' miracles being signs. They were never merely magical conjuring tricks but were always a purposeful manifestation of the spiritual power associated with God's Son.

Another thread is that of God's providential care for his creation. In Luke 12 we are told that God shows concern for sparrows (verse 6), ravens (verse 24) and lilies (verses 27-28). And if so, how much more will God faithfully provide for those who follow him! We need be anxious for nothing!

We are told to pray for others as well as ourselves. In his teaching on prayer, Jesus promised: 'Ask, and it will be given you; seek, and you will find; knock, and it will be opened unto you' (Lk 11:9). God will answer our prayer.

And so we see that Jesus recognised there was a real problem of suffering and evil, but at the same time he performed miracles, taught that God was utterly dependable in his providence and assured his disciples that their prayers would be answered.

Over the years, Christians have come to accept there is a profound mystery in all of this. They may not fully understand it, but many have found that in the suffering they are facing, God is there with them. The very fact of the life and death of Jesus shows that God does not stand aloof on the sidelines–he is involved in all the world's problems. Jesus was Immanuel, God with us. When he died on the cross, 'God was in Christ reconciling the world to himself'. It is the experience of many Christians that God is involved in their struggles and suffering that provides the practical solution to the deep theological mystery of God's involvement with his world.

But does modern science throw any light on the intellectual problems that remain? I believe that although there is no complete answer, there are hints that can illuminate our thinking. Naturally, at this stage the approach can only be provisional and speculative. It cannot be deduced directly from Scripture since modern science was not available to the authors of Scripture who lived in biblical times.

An evolving universe

We start by imagining (as far as we are able) our almighty, infinitely loving, omniscient God before the creation of the universe. He is a spiritual God, unconstrained in any way. But he wishes to create a universe which will contain creatures who will be sufficiently developed to have the capacity to respond to his love. This means that if he is to achieve this, they must be given freewill in order that they can choose to respond. It also implies that they will be free to rebel. This does not imply that God is excluded from influencing the course of events. The Persian, Cyrus, was probably quite unaware that he was fulfilling God's purpose when he defeated the Babylonians. God can influence events in this way even if he very often relies on conscious human collaboration. We recognise from our own experience that we do have moral and spiritual choices to make and have a degree of freedom in so doing. God has created for us the possibility of sinning but longs that we should not do so. And we know that over the centuries humankind has become caught up in a web of sinfulness from which we cannot, on our own, extricate ourselves. This, surely, is the nature of original sin which has spoilt what was intended to be a close relationship between God and humankind. And it is as a result of human sinfulness that so much suffering occurs in the world.

But what about natural disasters? Why should our loving God permit these to happen? Perhaps the universe in which we live is so special that there was no real alternative. We saw in Chapter 3 how finely tuned the universe had to be for it to be sufficiently long lived for advanced life forms such as ourselves to develop. We also saw that the earth is made up of the radioactive waste from stars that exploded. This process gives us the ingredients such as carbon and oxygen that are needed if life is to evolve. But it also gives us radiation which can both give rise to the mutations that are necessary for evolution to proceed and also give rise to cancer. The stresses in the earth's crust as it moves slowly under the influence of this radioactivity give rise to earthquakes. The life-giving energy from the sun is the same energy that gives rise to hurricanes and floods. And the process of evolution which seems so profligate and cruel may have been the only practicable way of enabling humankind to emerge given the physical constraints that must apply if he is to fulfil the role that God has given him.

We have seen that the nature of science is extremely subtle. There is a high degree of regularity and predictability in the laws of nature, God's laws. Humankind needs to live in a rational, dependable environment. But there is also a degree of unpredictability (quantum theory and chaos theory) that is necessary for creativity and freedom of choice. Natural suffering seems to arise inevitably from this mixture of randomness and determinism that is essential if God's objective of enabling humankind to evolve is to be realised.

God's action and scientific law

We have already seen that miracles are rare events about which, by definition, science can say nothing. But what about God's day-to-day providential care and guidance? Does God have to go against the rules of science (his rules!) in order to do this? We shall argue that this is not necessarily the case.

An analogy that some might find helpful is that of a game of chess. But, like all analogies, it must not be taken too far. If some intelligent observer who had never heard of chess, was shown a number of games being played, he would be able to deduce the basic rules. He would rapidly discover that pawns moved up the board, bishops moved diagonally, and so on. But, being intelligent, he would deduce that there was more to it than merely a set of rules. There was a purpose in the game–to capture the opposing King–and the strategies involved in the process were very complex and subtle. The basic rules were simple and never violated and yet, there was sufficient openness or freedom, that a sophisticated and intellectually challenging game was possible. The essence of the game is to be found in the battle that is raging rather than merely in the rules alone, although they are never violated.

Now we look again at God's interaction with his universe. There are 'rules'. We observe them as the laws of physics, laws which we now know contain enough flexibility for subtle and complex interactions to be feasible. Moreover, in spite of some difficult philosophical problems, most people would accept the commonsense view that we possess free will. The future is determined, at least in part, by our own personal decisions. We may not, as yet, completely understand how this works, but this in no way obviates the reality of free will.

And if we can influence the world, how much more can

God? He upholds and sustains it. But, just as the players in the game of chess influence the shape of the game without breaking the basic rules, so God works in his universe without normally going against his laws of science. Christians can look at the universe from a scientific perspective, using all the powerful tools modern science makes available to them, and find a beautiful and subtle pattern. But they can, with equal validity, examine the world with the eye of faith and discover meaning and purpose. They believe God has a plan for their lives and discover this as it unfolds. They find themselves presented with opportunities to serve God and help others that seem to have arisen by more than mere coincidence. They discover a hidden power available within themselves when faced with situations that they would have thought were beyond them.

There is a hard-won scientific route into the understanding of reality. Equally there is also a spiritual route which is not easy. The scientist spends countless hours with his apparatus and countless hours pondering the significance of what he has found. The Christian grows spiritually by spending time to allow God's Spirit to fill his or her soul and direct the mind. Above all it requires a personal costly commitment of obedience to the received word.

Knowing the rules of chess is not sufficient to enable an individual to play the game well. It is a necessary foundation on which the player's skill can grow. Similarly, knowing the laws of physics or biology is not sufficient to enable us to understand the universe fully. It is only when we recognise the spiritual dimension that God's universe is revealed in all its glory.

Divine and human action

We have seen that the nature of reality as revealed by science consists of an ordered pattern but with areas of uncertainty that, from their very nature, cannot be penetrated by science. It seems likely that, as individuals exercising our own wills, we are able to influence the physical world by mental processes which transcend the physical world. And if we as human beings can do this, it is reasonable to postulate that God can influence his creation.

God (and we) are able to influence the physical world by spiritual means which are outside the scope of science to investigate and yet the effect does not normally contravene the laws of science. It follows that although God is active in his universe, we should not expect to be able to discern that activity by scientific methods. God could, of course, compel us to respond by an unambiguous and overwhelming demonstration of his power. But his method is to win us by his love.

But this does not mean that God's activity is any the less real for all that. Note first of all that we are not confining God's activity by the traditional 'God of the Gaps' approach. This argued that God could only act freely in those areas where scientific understanding had not yet penetrated. As science advanced, God appeared to become more and more constrained. Our approach is different in that it highlights the basic unpredictability that has emerged from the discoveries of modern science. And although the extent of this unpredictability at any point in time may seem small, Chaos Theory tells us that the ultimate effect can be vast.

So we are not meant to see God's activity other than through the eye of faith. It is easy to see how it can be hidden from any scientific view. We observe the hurricane – but we cannot trace its cause back to the individual butterfly flapping its wing. If this is true of science, it is true of our experience of life as a whole. For example, for those of us who are married, our lives will have been greatly influenced by our choice of spouse. And many couples, looking back, would point to the apparently coincidental way they first met each other. A whole future hanging on a seemingly random event.

Or consider the dinosaurs. Sixty-five million years ago they dominated life on earth but became extinct leaving a niche for the mammals to occupy, eventually leading to the

evolution of humankind. Many consider the cause of this to be the massive environmental change resulting from a chance collision of a meteorite with the earth. Was this an accident or was it God's way of ensuring the creatures capable of responding to his love would evolve? It is impossible to prove one way or the other scientifically. One's view will be influenced by whether one believes we are merely the products of blind chance or children of a loving, active God.

The fact that God's activity is not discernible by the methods of science does not mean there is no pattern to it. We should expect, with the eye of faith, to recognise activity that is consistent with God's nature and his purposes. He will be concerned with justice, righteousness and love. At the same time he will continue to respect the free will he has given to humankind. This means that God has deliberately chosen to make himself (and therefore his creation) vulnerable to our wills and activities. We have been given the power to enhance or impede God's activity. It is probably no accident that in the Bible miracles have been most prevalent at times of spiritual advance and awareness. God's power is most manifest when his people are expectantly waiting to receive it and are willing to obediently co-operate with his leading. That is why prayer is so powerful. It is not us persuading God to do something he would rather not do. It is a case of us being open to understanding God's will and responding in obedience to the issues that are of concern to God. 'Whatever you ask in my name, I will do it, that the Father may be glorified in the Son; if you ask anything in my name, I will do it' (Jn 14:14).

If we can advance God's purposes by active co-operation we can impede them by our blindness, insensitivity, lack of concern and sinfulness. This is surely the meaning of the quotation from Isaiah at the beginning of this Chapter: 'Behold, the Lord's hand is not shortened, that it cannot save, or his ear dull, that it cannot hear; but your iniquities have made a separation between you and your God, and

your sins have hid his face from you so that he does not hear'
(Is 59:1-2).

God will work wonders when we repent of pride and
self-sufficiency and in humility align our wills to his perfect
will.

Summing up

We have seen from earlier chapters that our old-fashioned
view of a deterministic universe is no longer tenable.
Although we can thank God that much of what we experience is regular and dependable, modern science reveals the
significance of random and chance elements which enable a
universe of great subtlety and fruitfulness to have emerged.
It is in this universe, created by him, that God's purposes can
be worked out. At the centre of this purpose is the ability of
humans to respond by free will to God's love. This response
is an act of will lying outside the realm capable of being
investigated by science. And yet the visible consequences do
not violate the normal laws of science. The same is true of
God's actions in the world. These can only be discerned by
the eye of faith but are nevertheless real and powerful.

To the sceptic they will appear to be chance events or
coincidences. The person of faith will identify them because
they reflect the nature of God and reflect his purposes of love
and justice.

7

The Bottom Line

The world is charged with the grandeur of God....
Because the Holy Ghost over the bent
　World broods with warm breast and with ah! bright wings.

'God's Grandeur', Gerard Manley Hopkins

WE HAVE COMPLETED OUR SURVEY of modern science and its relationship to Christianity and we must now attempt to draw the threads together. What are the critical issues? What does it mean? In modern parlance, what is the bottom line?

God is back in the picture

Perhaps the first point to note is that scientists are, once again, talking openly about God in their description of science. It will be recalled that some two hundred years ago, Laplace was telling Napoleon Bonaparte that when it came to the role of God in his science, he had 'no need of that hypothesis'. Now we must not make too much of this point, but many eminent scientists, in explaining their work, find it helpful to talk about God–even though they may not be traditional believers. Einstein was fond of saying 'God is a mathematician' or, when debating with Niels Bohr, he wrote, 'Does God play dice?' It can be argued that Einstein was using the term 'God' as merely a shorthand for a first cause or the ultimate nature of reality. But he was implicitly accepting that there is a logical pattern to the reality that science is attempting to explore. Stephen Hawking, who would certainly not regard himself as a committed Christian, closed his popular best-seller *A Brief History of Time* by referring to a possible complete physical theory and its relevance to the question 'of why it is that we and the universe exist. If we find the answer to that, it would be the ultimate triumph of human reason—for then we would know the mind of God'. Once again we must not attempt to draw too much from this. It reveals a very narrow view of God. But Hawking himself recognises the relevance of his work to the role of the Creator.

This use of the word 'God' is not restricted solely to physicists talking about the origin and physical behaviour of the universe. At the other end of the scientific spectrum is psychology. Perhaps the most influential psychiatrist after Freud was Carl Jung. Some time ago I watched a repeat of an old television interview of Jung by John Freeman. Jung was, by then, an old man nearing the end of his life. Freeman's final question to him was whether, after a long career in psychiatry, it was possible for Jung to hold a belief in God. Jung drew himself up, banged the floor with his stick and affirmed, 'I know! I know!' Now, once again, Jung's view of God may not coincide precisely with the traditional Christian view. But the point is that, when eminent scientists find it natural to talk about the role of God in their work, then we Christians should sit up and take notice.

But it is not just talk. The very nature of the universe which modern science reveals, encourages a belief in God. It does not prove God exists in any strict, logical sense. But the balance certainly tilts in favour of a reasoned faith. And this point is highlighted by the now general recognition by scientists that the universe in which we live is a very special place—human life as we know it only being possible because of a series of remarkable apparent coincidences.

A very special universe

Humankind has evolved at the end of a long and complex process, starting with the creation of the universe in the Big

Bang, the formation of galaxies and stars, the production of vital chemical elements inside early stars, the condensation of those elements to form a planet with a suitable climate for life to evolve, the planet circling a star that is sufficiently long lived for there to be time for advanced life-forms to develop, and finally the transition from inanimate to replicating life-forms to initiate and to develop.

Each step in this chain seems improbable–some incredibly unlikely. For example, let us consider the initial Big Bang. We saw in Chapter 3 that, for the universe to have lasted sufficiently long for galaxies to have formed and life to have evolved, the speed of expansion must have been just right. At one second after the Big Bang the margin for error was only one part in 10^{15}, an incredibly small figure. For those who play golf, it is the probability of going out and holing in one in each of the first three holes.

Now there are basically three ways of accounting for this degree of fine tuning.

1) God, the super designer, got it just right.

2) There is something in the laws of physics that, if only we understood it properly, would account for this remarkable precision. The inflation theories which postulate a very rapid expansion of the universe in the first split second are an attempt at just such an explanation. But even if they are correct, it leaves the question of why the laws of physics are such that the theory can have validity. In other words, God organised the laws of nature so that it was bound to happen. It is another version of 1)!

3) We postulate that there are a vast number of universes that we have no way of knowing anything about. We happen to live in the one in which the conditions happened to be just right for human life to evolve. It is rather like saying that, if people played golf for a sufficient length of time, someone would eventually hole in one in each of the first three holes in succession. And this is the game that everyone would remember.

We also saw in Chapter 3 that in order for carbon and heavier elements to be produced inside stars the nuclei of carbon must have certain properties that lie within closely defined limits. Life as we know it could not have evolved without carbon. So there is a further 'coincidence'. It is as though our golfer, having holed in one in each of the first three holes, goes back to the club house and wins the jackpot on the fruit machine!

Then there is the issue of why the laws of nature are such that planets such as the earth can exist with conditions suitable for life to develop. Professor Sir Martin Rees has worked out what would happen if gravity were stronger than it is, all other physical effects remaining the same. Gravity is a very weak force. The repulsive force between two protons due to their electrical charge is 10^{38} times as big as the attractive force between them due to gravity. Rees worked out what would happen if this factor was 10^{28}. He found that the stars in this universe would only be about 2 kilometres in diameter and would burn out in about a year. Only minute organisms could withstand the massive gravitational forces on any planet surrounding such a star and it is hard to see how advanced forms of life could have developed, even if they had the time to do so in the year's life of their solar system.

And so we understand why scientists have come to recognise that the universe is a very special place. Professor Paul Davies, by no means a Christian believer, recently wrote: 'Through my scientific work I have come to believe more and more strongly that the physical universe is put together with an ingenuity so astounding that I cannot accept it merely as brute fact.'

Science cannot answer the question of why it should be so special. Do we just happen to live in the one of a vast number of universes where conditions were just right for life to evolve? Or is there a purposeful designer behind it all? The Christian view of a creator God seems highly plausible.

Demise of determinism

It will be recalled that for the two and a half centuries following Newton, it seemed that science was suggesting that we lived in a deterministic universe. The future was fixed by what was happening in the present. There was no room for free will. There was apparently no scope for God to act in his universe.

All this has changed with the discovery this century of quantum mechanics and the understanding of the behaviour of complex systems that has been gained (chaos theory). Scientists now see uncertainty and chance lying at the centre of their understanding of reality. There is now scope for the responsible exercise of free will without violating any scientific laws. Equally, if one holds the Christian view that God is the living Lord who is active in his world, then one can legitimately claim that such a view is not contradicted by any scientific law or discovery.

Moreover, we have come to appreciate that the world that is revealed by science is one that contains both order and chance and this is just the type of mixture that is likely to lead to fruitful developments in complex systems. A small change in one part of such a system can lead to a significant effect elsewhere.

The popular example that is often quoted is that of the butterfly flapping its wings in one part of the world and thereby initiating a hurricane that develops thousands of miles away. It is in just such a universe that one can envisage God quietly influencing and steering events with a gentle but firm hand on the tiller. Christians can rejoice that they too have been given free will which can be used to enhance God's purposes even if, in practice, we know that we often act in such a way as to obstruct them.

The problem of good

This process of fruitful development is clearly seen in the evolution of the universe, starting with the Big Bang and culminating within the last one thousandth of one percent of time in the emergence of that highest form of life, humankind. We presented some views of the nature of man in Chapter 5. In one sense he can be seen as the end product of a long process involving random mutations by which living organisms evolve into more advanced forms under the pressures of natural selection. Certainly there is overwhelming scientific evidence to support our biological continuity with other species. But this is not the whole story. Man is a very special creature in several ways. Here we just highlight one aspect – the issue of why we applaud goodness in people when we see it.

All of us have something inside us that recognises that we are moral creatures. Why is it that we instinctively applaud that which is unselfish and self-sacrificing in people when we see it? Our hearts are warmed by a Mother Teresa and chilled by a Hitler.

But how can these feelings have arisen if we are merely the product of random mutations and natural selection? Hitler, who appals us, in seeking to develop his master race was more in tune with the survival of the fittest than Mother Teresa. Where is the genetic advantage in an elderly spinster caring for the dying in the streets of Calcutta? We surely know that we are more than merely the product of evolution. It may be true to say there is a biological continuity between ourselves and other primates but our description of man would be incomplete without recognising that we are very special creatures, made in the image of God.

How reliable is scientific knowledge?

In recent years there has been a growing anti-science movement. It seems to have arisen, in part, as a symptom of a discontentment with our post-Enlightenment society. In part it is due to concern with the issues that have arisen as a result of the widespread application of science in the second half of the twentieth century. But it is partly a result of a reaction to a claim by some scientists, which has gained a degree of popular acceptance, that science is the only source of reliable knowledge.

The issue was parodied clearly if somewhat cruelly by Michael Flanders in the review *At the Drop of a Hat*: 'One of the problems of the world today is undoubtedly this problem of not being able to talk to scientists, because we don't understand science; they can't talk to us because they don't understand anything else, poor dears.'

Now this book has in no way adopted an anti-science stance. On the contrary, it is written in the firm conviction that God has created a rational universe and we can learn much of the nature of reality through scientific investigation. But at the end of the day, science will only take us part of the way. We can also discover meaning through history, philosophy, literature, poetry and art as well as, supremely, the revelation of God's nature and purposes in the life of Jesus Christ. That is why Stephen Hawking is so naive to suppose that, if we were able to develop a comprehensive mathematical theory of the early universe, then 'we would know the mind of God'. God is the great creator and sustainer of the universe. But his mind is concerned with love, justice, holiness and righteousness just as much as with physical science.

Accepting, then, that science provides a powerful, if partial, means of viewing the nature of reality, are there any limitations to the scientific method itself? There would appear to be three points that are of most significance.

The first is that scientific theories are always provisional. They are only valid so long as they are supported by all the available evidence. Fresh evidence might require the theory to be modified or even completely overturned. For example, Newton's laws were found in this century to be inadequate to account for the behaviour of very small particles. The new quantum theory, which approximates to Newton's for large bodies, has been extremely fruitful in illuminating a broad and perhaps unexpected range of science from chemistry to the behaviour of solids, radioactivity, spectroscopy and nuclear physics. Science is progressing most rapidly when its accepted wisdom is being overthrown. In the late nineteenth century, the American physicist, Michelson, expressed the view that the fundamental problems of physics had all been solved. Within a few years he had carried out a famous experiment with Morley that shattered this confidence and was a significant factor in leading Einstein to develop the new theory of relativity. Then again, in the light of the new understanding of physics that came in the late 1920s in the wake of quantum mechanics, the eminent physicist Max Born said he believed that all significant physics problems would be resolved within six months! This was just before Chadwick discovered the neutron, heralding the modern elementary particle era!

Secondly, scientific theories are most authoritative when describing events within the range of experimental observation. This is why Newtonian mechanics was so successful for so long. It was only when scientists began to consider extremely massive or fast-moving objects that one set of weaknesses became apparent—weaknesses that could be remedied by relativity theory. Another set of weaknesses became apparent when scientists began to experiment with very small particles—weaknesses that could be remedied by quantum theory. Scientific theories are also most firmly based when they are supported by a wealth of experiments that have been repeated in separate laboratories. Clearly, accounts of the origin of the universe or the origin of life on earth are not in this category. Both from the one-off nature of each event and from the considerable extrapolations involved from the established range of scientific data, they cannot be. This does not mean that these theories are wrong. Of course not. What it does mean is that there will inevitably be a degree of speculation in the accounts which it is important to recognise.

The third weakness is really a logical trap into which some scientists seem to have a tendency to fall. We met it in Chapter 2. It is reductionism. Now a very powerful method of understanding, for example, the biological subject of genetics, is to focus on the chemical and physical processes that cause our genes to act in the way they do. Hence the significance of molecular biology which has revolutionised our understanding of genetics. This form of reductionism is perfectly acceptable and forms a powerful scientific approach. What is wrong is to reverse the argument and claim that biology is nothing but physics and chemistry. The fallacy in this approach can be immediately appreciated if one imagines applying it to an aircraft that one has come across for the first time. One could observe that there were considerable quantities of metal and one could study these properties. Then there were jet engines–and one could study the fluid dynamics of these. There were radios and navigation equipment. But however far one got with studying their individual aspects, one would not be in a position to design an aircraft that would fly unless one also had an understanding of the specifically aeronautical whole, centred on aerodynamics. In just the same way, those who claim that a complete knowledge of the whole of science can be obtained from a knowledge of, say, physics alone are deluding themselves.

And so we see that science is capable of providing valuable knowledge about God's world. But it is not complete. God is greater than his creation and cannot be contained by our necessarily limited scientific vision. God reveals himself through his natural world–but a deeper understanding of God and his purposes is to be obtained by other means which will complement and not contradict science. Above all we can see the nature of God most clearly when we ponder the life and death of Jesus: 'He is the image of the invisible God, the first-born of all creation; for in him all things were created, in heaven and on earth, visible and invisible, whether thrones or dominions or principalities or authorities–all things were created through him and for him. He is before all things, and in him all things hold together' (Col 1:15-17).

'God is a mathematician'

This is a famous phrase of Einstein's. It reflects the fact that, as far as we can judge, the physical universe can be modelled in mathematical terms. This, in turn, implies that there is a logical structure to it that, as scientists, we seek to comprehend.

Moreover, scientists who have made significant discoveries, have often been struck by the elegance, simplicity and beauty of what is revealed. No wonder Archimedes shouted 'Eureka' and leapt out of his bath when he, for the first time, understood the simplicity and elegance of how to calculate buoyancy. And this sense of wonder has been experienced by the great scientists down the ages.

So what are we to say of the universe? The Christian view is that there is a purpose and a design behind it. We live in a Designer Universe. Surely this view is in tune with the general picture revealed by science.

An alternative view is that we are merely the product of chance. We just happen to live in the one of an almost infinite number of universes where conditions were just right for life to develop. And this development then proceeded by a series of random steps, the successful outcomes being selected solely by the survival of the fittest. What a despairing scenario! It accounts for the worst features of human nature but can say nothing of man at his highest. What purpose and meaning can there be in such a world? And why should science be possible? Why should there be a pattern to be discovered if it is all so chancey?

Science, as we know it, took off in Western Europe during and after the Reformation. It could potentially have taken off earlier in Greece, Egypt, China or in Arab countries. It is impossible to prove why this should be but many historians consider that Christian belief in an orderly world created and ruled by a single omnipotent God was the decisive factor.

For too long Christians have been on the defensive about the discoveries of science. Certainly, over the years, these discoveries have caused considerable fundamental thought about the relationship between the two. This book has attempted to demonstrate that modern scientific discoveries have revealed a universe that is at one with Christian revelation.

'Charged with the glory of God'

Once we accept that there is no fundamental conflict between science and Christianity, we can relax and look at scientific revelation with fresh and wondering eyes. We can turn to worship our great God who has created and sustains such a remarkable universe.

The danger of science on its own is that it is limited by human ability to observe and to reason. In that sense it is man centred. Bring God into the picture and we obtain an added dimension which provides a depth that cannot be appreciated by science alone. But God does not come in as an 'add on'. His is the primary role. He is the creator, we are the creatures. He is the source of meaning and purpose–we can only seek to understand. His is the plan that led to the very special universe in which life could evolve–we can only marvel at this. His is the hand that guides the evolutionary process leading, after four and a half billion years, to the emergence of creatures able to recognise and respond to him–we can only feel a sense of awe and wonder. He has left an openness in the structure of nature so we have free will–we can only use this free will to glorify him. He operates in his universe, caring for each one of us–we can only thank him for his goodness and seek his will through prayer. Wonder of wonders, we see all the loving nature of God revealed in Jesus Christ. We can only respond in thankfulness, trust, obedience and worship.

Just over a hundred years ago, Gerard Manley Hopkins, the Jesuit poet, wrote, 'The world is charged with the grandeur of God.' It is both through Christianity and through science that we see something of this grandeur of the great God whom we worship and serve.

APPENDIX 1

UNPREDICTABILITY OF
THE FUTURE

W E HAVE SEEN FROM the main text that scientists of the eighteenth and nineteenth centuries such as Laplace considered that, from a detailed knowledge of the universe at one moment in time, it is in principle at least possible to predict its state at all future times. This Appendix presents a simple calculation to show how rapidly this breaks down.

Consider the molecule in air at atmospheric pressure and room temperature. Air consists mainly of nitrogen and oxygen molecules travelling at a speed of about 450 m.sec^{-1} and colliding with other molecules every 10^{-10} seconds or so. Let us focus on one molecule and ask how accurately we can predict its position after merely 10^{-10} seconds–ie, the time between one collision and the next.

Laplace would have considered that it was possible to know, at least in principle, the position and velocity of every molecule and hence deduce precisely where they were 10^{-10} seconds later. But twentieth-century quantum theory shows there is a limit to this precision. You cannot know exactly both the position and velocity of a particle. Heisenberg's Uncertainty Principle puts the matter mathematically as:

$\Delta x \Delta(mv) > \hbar$
Δx is the uncertainty in position
Δv is the uncertainty in velocity
m is the mass of the particle
$\hbar$ is a constant, known as Planck's Constant divided by 2π.

Now, let us consider the uncertainty in the position of the molecule at right angles to the direction of travel. Initially there will be an uncertainty in position Δx_o and speed Δv. After time t, this uncertainty in position will have grown to:

$$\Delta x_o + t\Delta v$$

Using Heisenberg's Uncertainty Principle, this must be at least:

$$\Delta x_o + \hbar t/m\Delta x_o$$

Now this is of the form $ay + b/y$ and one can find by simple calculus that the minimum value is $2\sqrt{ab}$.

Hence the uncertainty in the position of the molecule after time t is at least:

$$2\sqrt{\hbar t/m}$$

If we put:

$\hbar = 10^{-34}$J.s.
$t = 10^{-10}$s.
$m = 5 \times 10^{-26}$kg

we find an uncertainty in position of about 10^{-9}m.

This is three times greater than the diameter of an air molecule which is typically around 3×10^{-10}m.

Hence there is no way of taking an air molecule and predicting which other air molecule it will collide with next.

APPENDIX 2

CHAOS THEORY

ALTHOUGH MANY OF the basic ideas were formulated by the French mathematician, Henri Poincaré in 1890, it is only within the last twenty years that the wide implications of chaos theory have become appreciated. It will be recalled, from the main text, that so-called chaos ensues when only small changes in the initial conditions fed into the mathematical equations describing the behaviour of a system lead to large changes in later behaviour. The unpredictable behaviour of the motion of a ball-point pen stood vertically on its point is a very simple example. The way in which it falls will be determined by very small changes in the initial motion. The surprising feature of this behaviour is that it occurs widely in nature. And secondly, unexpected patterns of behaviour can often arise.

In this Appendix we shall look at a particularly simple equation to develop an understanding of how complex behaviour can arise and to illustrate some of the more elementary mathematical approaches to the subject. Then we will look at some examples of where chaotic behaviour occurs in practice.

A simple equation

Consider an environment capable of supporting a certain number of animals but no more. Suppose at a given point in time there is a fraction x of this maximum number. Clearly, x can lie between 0 and 1. Suppose at one point in time x takes the value x_n. Now consider the value of x one generation later. Simple breeding would suggest the value would increase to kx_n, but if x approached unity, this figure would have to be reduced to reflect the shortage of food. Let us represent this approximately by introducing a factor $(1-x_n)$. Our simple model is then:

$$x_{n+1} = kx_n (1-x_n) \qquad - (1)$$

It is the properties of this equation that we are going to study.

Note that $x(1-x)$ has a maximum value of 1/4 so k must be less than 4 if x is always less than 1. Also k must be greater than 1 otherwise the line would rapidly die out.

Now the perceptive reader would note that there is a simple solution to equation 1. That is x_n is a simple constant value for all n. We can find this simply by writing:

$$x = kx(1-x)$$

which has the solution,

$$x = (k-1)/k \qquad - (2)$$

Is not this the end of the story? Would not the system always settle down to this value?

Let us check. Suppose we choose k=2 and start with $x_1 = 0.3$. According to (2) we would expect x to settle down to a value 0.5. By substituting in (1) we find:

$x_1 = 0.3$
$x_2 = 0.42$
$x_3 = 0.4872$
$x_4 = 0.49967$
$x_5 = 0.4999997$

Notice how rapidly x converges to 0.5. It is as we would expect a well-behaved system to behave.

Now let us choose a value of k of 3.6. According to (2)

we expect x to settle down to a value of 2.6/3.6 or 0.72222.... Let us see what happens when we start, as before, with $x_1 = 0.3$. We get:

$x_1 = 0.3$
$x_2 = 0.756$
$x_3 = 0.6640704$
$x_4 = 0.8030912$
$x_5 = 0.5692885$

We see that this time the numbers are all over the place–we have chaos–or, at the least some complex behaviour..

Let us try again. This time we keep k = 3.6 and start with $x_1 = 0.722$, very close to the value from equation (2) of 0.72222.... This time we get:

$x_1 = 0.722$
$x_2 = 0.7225776$
$x_3 = 0.7216531$
$x_4 = 0.7231316$
$x_5 = 0.7207642$

Note how the succeeding values of x depart, increasingly from the value of 0.72222...predicted by equation (2). Instead of converging on this value it diverges from it just as the ball-point pen does when it falls from a vertical position. We are in an unstable state for this value of k (3.6) and it is this that leads to the chaotic behaviour.

Limit of regular behaviour

This example gives us a clue to discovering the limits of applicability of equation (2). Let us suppose $x_1 = (k-1)/k + \varepsilon$ where ε is a very small number.

From equation (1) we find:

$$x_2 = k \left\{ \frac{k-1}{k} + \varepsilon \right\} \left\{ 1 - \frac{k-1}{k} - \varepsilon \right\}$$

This simplifies to:

$$x_2 = \frac{k-1}{k} + (2-k)\varepsilon - k\varepsilon^2$$

We can ignore the term in ε^2 since ε is small. The divergence from $(k-1)/k$ has changed from:

ε in x, to $(2-k)\varepsilon$ in x_2

If k lies between 1 and 2, the divergence in x_2 is less than in x_1 and the series converges.

If k lies between 2 and 3, the divergence in x_2 is less than in x, but the sign is different. Once again the series converges.

If k is greater than 3 the sequence diverges as we saw with the example k = 3.6 calculated above.

So we expect complex and possibly chaotic behaviour for values of k greater than 3.

New, unexpected patterns

Now let us look at what happens when k is just greater than 3. It is possible to show by computation (it is tedious without a computer) that the value of x can oscillate in a regular way.

Let us suppose $x_1 = x$
Then $x_2 = kx(1-x)$
               $x_3 = k.kx(1-x)\{1-kx(1-x)\}$

Now look for solutions where

$x_3 = x_1 = x$

We have:

$x = k.kx(1-x)\{1-kx(1-x)\}$

or

$k^3x^3 - 2k^3x^2 + k^2(k+1)x + 1 - k^2 = 0$

This is a cubic equation in x. The solutions are:

$$x = (k-1)/k$$
$$x = \{k+1+\sqrt{(k+1)(k-3)}\}/2k$$

and

$$x = \{k+1-\sqrt{(k+1)(k-3)}\}/2k$$

The first of these solutions is the familiar one of equation (2). It appears since $x_1 = x_2 = x_3$ is a solution of $x_1 = x_3$.

The second and third solutions provide the two values of x between which the oscillation takes place. Note that they only come in for values of k greater than 3 since for lower values the number in the square root would be negative. Also at k = 3, the three solutions are all equal (x = 2/3). Although the solution for stable x changes character at k = 3, there is no discontinuity.

So we see that at k = 3 and above, there is a stable oscillation between two values. But this does not apply to all values of k. As k increases to 3.45, the period changes to 4; at 3.56 it doubles again to 8 and so on. Beyond 3.57 the behaviour goes chaotic but there are regions between 3.57 and 4 where regular patterns emerge. For example, for values of k around 3.835 there are oscillations with a period of three.

So we have seen that for the simple equation:

$$x_{n+1} = kx_n (1-x_n)$$

The behaviour varies dramatically depending on the value of k that is chosen. For k between 1 and 3, the value of x tends to settle to a steady level. Between 3 and 3.57 there are oscillations, the period of which depends on the precise value of k. Between k = 3.57 and 4, the behaviour is generally chaotic but there are small regions where periodic behaviour is obtained.

Very often the real world is not simple. And yet, within this very complexity, unexpected patterns of behaviour can emerge. This was true of our simple equation–it is true in many areas.

Chaotic behaviour is observed in many real situations. Whereas the motion of the planets in the solar system is perfectly stable, the motion of, say, a speck of dust in a universe containing only it and two large bodies is extremely complex. This was the system originally studied by Poincaré.

It is well known that, at low speeds, the flow of a fluid in a tube is regular but, at some critical speed the flow suddenly becomes highly turbulent. Is this an example of chaos setting in? Many flow situations demonstrate a periodic behaviour. An example is the regular vortices that are set up as fluid flows over a cylinder–this is why we hear the wind whistle in overhead lines.

We have already seen how complex and sometimes chaotic behaviour can arise in population dynamics. The existence of years of plagues has been known from times of early history. It is only relatively recently that it has been possible to take a proper analysis of this phenomenon into account.

We end this review of chaos by mentioning dynamic chemical oscillations. The effect was apparently first noted by William Bray in 1921 but became well known in 1958 after the Russian chemist, B.P. Belousov, observed oscillations in the colour when a number of salts were mixed together. The work was extended by Zhabotinskii who showed that waves of colour could form, if the chemical mixture is spread in a thin layer. This is an example of the complex behaviour that can occur when operating far from chemical equilibrium. There have been suggestions that such processes could be important in biological development but this appears to be an interesting suggestion rather than hard fact.

APPENDIX 3

QUANTUM THEORY

Q UANTUM THEORY HAS BEEN the most influential development in physics this century, providing a foundation for much of modern physics. In this appendix we look at one simple example, our old friend the ball-point pen, before looking a little deeper into the theoretical basis of quantum theory.

Ball-point pen

It is often said that quantum theory applies to very small particles–on the atomic scale and below–but that Newtonian mechanics applies to normal-sized objects (like ball-point pens!). This is normally true, but in an unstable situation (like a ball-point pen stood on end) a quantum fluctuation could be sufficient to initiate the motion.

Let us represent a ball-point pen by a mass m half way up a light rigid bar of length 2r. We stand the bar vertically and calculate how long it takes to get moving. Once it starts moving through only a small angle it will rapidly fall over (see top of following page).

The equation of motion is:

$$mr \frac{d^2\theta}{dt^2} = mg \sin \theta$$

For small values of θ, $\sin \theta \sim \theta$. Hence:

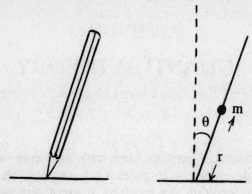

$$mr \frac{d^2\theta}{dt^2} = mg\,\theta$$

$$\theta = Ae^{\sqrt{g/r}\,t} + Be^{-\sqrt{g/r}\,t}$$

Suppose at t = 0, the centre of gravity is moving at a speed v and is a distance x from the vertical. Both quantities are small.

Hence, at t = 0

$$\theta = x/r \text{ and } \frac{d\theta}{dt} = v/r$$

Hence

$$A + B = x/r$$
$$\sqrt{g/r}\,(A - B) = v/r$$

So

$$A = \tfrac{1}{2}\,(x/r + v/\sqrt{gr})$$
$$B = \tfrac{1}{2}\,(x/r - v/\sqrt{gr})$$

And

$$\theta = \tfrac{1}{2}\,(x/r + v/\sqrt{gr})e^{\sqrt{g/r}\,t} + \tfrac{1}{2}\,(x/r - v/\sqrt{gr})e^{-\sqrt{g/r}\,t}$$

Now $\sqrt{g/r}$ is about 10 s.$^{-1}$, so the second term rapidly becomes negligible compared with the first. We may then write:

$$\theta = \tfrac{1}{2}\,(x/r + v/\sqrt{gr})e^{\sqrt{g/r}\,t}$$

Now we know that Heisenberg's Uncertainty Principle gives

$$x.mv \gtrsim \hbar$$

Now, just as in Appendix 1, we have an expression of the form

$$\alpha x + \beta v$$

subject to

$$xv \gtrsim \hbar/m$$

The minimum value of $\alpha x + \beta v$ is

$$2\sqrt{\alpha\beta\hbar/m}$$

Hence

$$\theta \gtrsim (\hbar^2/gr^3m^2)^{1/4}e^{\sqrt{g/r}\,t}$$

Taking

$$r = 0.07\text{m}$$
$$m = 8 \times 10^{-3}\text{ kg.}$$
$$\hbar = 10^{-34}\text{ J.s}$$
$$g = 9.8\text{ m.s}^{-2}$$

we get $\theta \gtrsim 4.6 \times 10^{-16}\,e^{11.8t}$

For $t = 2.6$ sec, the angle θ is about 10^{-2} radian.
For $t = 2.8$ sec, the angle is about 10^{-1} radian.

Now this has not been a precise calculation, but it is adequate to demonstrate that with a normal, unstable situation, quantum effects are sufficient to disturb the system and, in this example, lead to observable effects in seconds.

Quantum mystery

So far we have used only one aspect of quantum theory, the impossibility of knowing precisely both the position and speed of a particle. But now we must delve a little deeper into this strange quantum world.

Most text books start by considering particles passing through a screen containing two slits and being detected in some way at some distance beyond them:

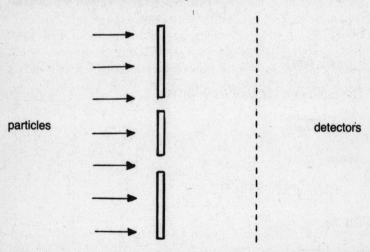

particles

detectors

Let us block up one of the slits, say the bottom one, and see what happens. The signals from the detectors show a spacial pattern as in the left hand picture below. The positions directly opposite the two slits are indicated (see top of following page).

It will be seen that the particles do not all land on the detectors in line with the open slit—there is spreading as

bottom slit closed   top slit closed

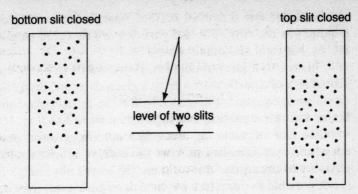

level of two slits

would be expected by Heisenberg's Uncertainty Principle. A simple pattern centred on the line of the lower slit is observed when the top slit is closed and the bottom one is open.

So far nothing very strange has happened. But what happens when both slits are open? Intuitively one would expect to see a superposition of the two patterns. Actually a series of bands appears:

It seems that the effect of opening a second slit in some way prevents some of the particles travelling through the other slit landing on certain regions of the detectors. But how can a particle passing through, say, the bottom slit possibly know the top slit is open and modify its behaviour accordingly?

This is the key issue that led to quantum mechanics. The

observed pattern is typical of that observed when waves impinge on the slits. But the particles are detected singly, one by one, and are unlike waves.

This led to a form of dualism. If one wants to know the probability of a particle arriving at a given detector, then one uses a wave equation to calculate it. But the actual behaviour at the detector is particle-like rather than wave-like.

One can calculate the wave-like behaviour quite precisely and such calculations form the basis of much modern physics and chemistry. But what are the waves physically? It is not possible to point to a physical description. All they do is determine the probability of a particle-like behaviour occurring at a specific time and a specific point in space.

Those who ponder the relationship between mind and brain and find the concept of some form of dualism difficult, might take comfort from the fact that there appears to be an equally perplexing dualism right at the heart of the physical world.

Do we understand quantum theory?

In one sense we do. The great majority of modern physics is based on the use of quantum theory. We know how to do calculations with it. But there is no complete agreement among scientists about the underlying basis.

The uncertainty all centres on exactly what happens when one observes the result of a process in which quantum theory has been involved. We saw with the slit experiment that the probability of a detector being activated by a 'particle' is determined by something with wave-like properties. But supposing the 'particle' emerges from the detector. Do we need a new set of waves to describe its subsequent behaviour? Or suppose we don't observe the detector until later (the detector could be a photographic plate that needs developing). Has the wave function changed before or after our knowledge of the event? Do you need a human observer for quantum mechanics to have any meaning? It is questions such as these that are hotly debated by quantum physicists.

APPENDIX 4

MATTER AND ANTIMATTER

B Y THE LATE 1920s it had become well established that all matter was composed of atoms which consisted of electrons (negatively charged particles) surrounding a positively charged nucleus. The behaviour of the electrons could be described by quantum theory. It was in a remarkable attempt to combine quantum theory with relativity theory that the Cambridge mathematician, Paul Dirac, was led to deduce that positively charged electrons or positrons should also exist. Three years later positrons were discovered in experimental studies of cosmic rays.

It became quickly understood that if a positron were to collide with a normal electron, then the two could annihilate each other with the emission of light. Conversely a light particle of high energy can be made to generate a pair (electron and positron) of electrons.

It is now recognised that each type of particle that exists in nature has its opposite counterpart. For any kind of matter there is the possibility of antimatter.

Why should our universe be composed mainly of positively charged nuclei and negatively charged electrons rather than the other way round? In other words, why do we live in a universe of matter and not antimatter? It seems that at one very early stage there was very nearly the same amount of matter and antimatter. But with the expansion and cooling of the universe, all the antimatter combined with matter with the production of photons, particles of light, leaving only the very slight excess of matter behind. That is why there are billions of photons in the universe for every atom in it. If it had not been for this very small asymmetry between matter and antimatter, then there would be no matter in the universe–only very faint electromagnetic radiation.

The standard model

By about 1950, considerable work had been undertaken on the structure and properties of atomic nuclei based on the assumption that they were composed of protons and neutrons held together by strong but unexplained forces. Then it was realised that it was possible to produce a large number of elementary particles using the powerful high-energy accelerators that were coming into operation.

Then, in 1962, Gell-Mann and, independently, Zweig, introduced the idea of quarks–particles with either one third or two thirds the charge of the proton. The quarks have another property known as colour which is a useful label but bears no relationship to the usual meaning of the word. By analogy with the calculation of the interactions between charged particles, which is known as quantum electrodynamics, the interaction between quarks is calculated using quantum chromodynamics. Quarks interact very powerfully with each other and never appear singly, only in threes. Whereas electromagnetic interactions are mediated by exchanging photons, the interactions between quarks are mediated by the exchange of particles known as gluons.

Although the concept of quarks was originally not generally accepted by the scientific community, there is now both substantial experimental and theoretical evidence to support it. The model described has become known as 'the standard model'. It has proved highly successful in predicting experimental results. Nevertheless, it is recognised that something better is required. In particular, the standard model does not include a proper theory of quantum gravity. Theoreticians are hoping to make progress in this direction by so called super-string theories. Such theories are at an early stage and certainly beyond the scope of this book.

APPENDIX 5

STEPHEN HAWKING'S MODEL OF THE EARLY UNIVERSE

E VERY SO OFTEN an individual captures the general public's imagination. Stephen Hawking is one such person. A Cambridge professor of mathematics who has made highly original contributions to the theory of cosmology, and a popular science writer who has produced a book that has so far sold 5 million copies worldwide, he has achieved all this despite having to operate from a wheelchair and with a voice synthesiser due to suffering from motor neurone disease which has been with him throughout his working life.

Hawking clearly recognises the significance of his work, which is concerned with the origin of the universe, to theology. We have seen in Chapter 3 that the universe is expanding and, if one calculates back in time, then the universe is predicted to get denser and denser. Ignoring quantum effects one would calculate that, some 15 billion years ago, it would have started at an infinite density. Our understanding of physics is not such that one can predict what would be happening at this infinite density point or singularity which some have identified with the moment of creation. Some would argue that it was at this point, when the physics broke down, that God had the opportunity to act.

Now Hawking has proposed a model of the early universe in which the density does not become infinite and in which the normal laws of physics can be expected to apply.

Although not all cosmologists agree with his approach and there are formidable difficulties—not least the fact that there is, as yet, no satisfactory theory of quantum gravity–it is not unreasonable to suppose quantum effects might have the effect of 'blurring' the high density early phase of the universe so that a singularity did not arise. Remember that the important time scale in all of this is just 10^{-43} seconds. After this time we are well away from the singularity.

He wrote:

> The idea that space and time may form a closed surface without boundary also has profound implications for the role of God in the affairs of the Universe. With the success of scientific theories in describing events, most people have come to believe that God allows the Universe to evolve according to a set of laws and does not intervene in the Universe to break these laws. However, the laws do not tell us what the Universe should have looked like when it started–it would still be up to God to wind up the clockwork and choose how to start it off. So long as the Universe had a beginning we could suppose it had a creator. But if the Universe is really completely self-contained, having no boundary or edge, it would have neither beginning nor end: it would simply be. What place, then, for a creator?.

How can we respond to this? There are two points to make. The first is that Hawking's science is right up to date but his theology is two hundred years out of date. Christians do not believe that God started up the universe and then left it to run its course without any interference or concern. That is eighteenth-century Deism not Christianity. Once we recognise that God is a living God, Lord of the universe, there is no point to answer.

Secondly, Hawking's model accepts that there are physical laws that applied just as much at the earliest moment in time as they do today. But he does not, and indeed cannot, answer the question of why the laws are as they are. Hawking himself recognises this point. He wrote:

In effect, we have redefined the task of science to be the discovery of laws that will enable us to predict events up to the limits set by the uncertainty principle. The question remains, however: How or why were the laws and the initial state of the Universe chosen?

There is a limit to how far we can go back in time. With no physical universe in existence, time has no meaning. This point was recognised by Augustine some 1,600 years ago but came into prominence with Einstein's Relativity Theory. If the physical laws had to be there for the physical universe to come into being, then these laws are in a sense eternal, created outside time, and science is powerless to probe their origin.

APPENDIX 6

FROM DNA TO PROTEINS

WE HAVE SEEN THAT our body is composed of cells, most of which contains a nucleus in which the genetic information controlling our growth and development is stored in large molecules known as DNA. Apart from errors due to mutations, the DNA in each cell is identical.

But DNA merely stores information. It is rather like a floppy disc that is fed into a computer. To be effective, this information needs to be read and used to build up suitable proteins in our cells. This Appendix describes how this is accomplished.

The building-blocks of proteins are amino acids of which there are twenty different types in our cells (and those of all living organisms). A protein is therefore a string of amino acids put together in a special order. So the problem is how to 'read' the information stored in the string of bases in the DNA and translate this into the construction of a string of amino acids to form the required protein.

The code in the DNA is simple. You read the bases off in groups of three. Then the first base could be either C, G, A or T—ie, there are four possibilities. Similarly there are four possibilities for each of the second and third base in the group. So there are $4 \times 4 \times 4 = 64$ ways of arranging the three bases, which is more than enough information to specify to which of the twenty types of amino acid it relates and, in addition, to have some arrangements of the bases meaning 'end of protein', etc.

The steps involved are:

1) The DNA partially unzips.

2) Inside the nucleus are bases of a molecule RNA which is somewhat similar to DNA but is single stranded and has the base thymine replaced by a similar one, uridine. A length of RNA is assembled corresponding to the particular gene on the DNA.

3) This length of RNA, known as messenger RNA, then passes from the nucleus to the main body of the cell (the cytoplasm). There the RNA becomes attached to the surface of a special structure known as a ribosome on which the protein will be assembled.

4) There are, circulating in the cytoplasm, amino acids, each being attached to a three-based section of RNA known as transfer RNA (tRNA). For each of the 64 possible tRNA base arrangements, only one type of amino acid will fit.

5) The tRNA amino acid pair arrive at the ribosome assembly line and the tRNA lines itself up with the complimentary mRNA. This puts the appropriate amino acid in the correct position.

6) The sequence of amino acids connect up to form the protein, which leaves the ribosome, and circulates in the cell.

The degree of integration in the whole of this process is remarkable. If we think of a growing embryo, information is somehow passed from the body as a whole (such as cell x becomes part of a leg or cell y becomes a blood cell). This information is, in some way, used to activate or switch off sections of DNA in each cell so just the right amount of the appropriate proteins are produced to build up the specific cell. This process, in turn, requires there to be special protein and RNA to be present which have themselves been generated by an earlier decoding process.

INDEX

SELECT BIBLIOGRAPHY

J.D. Barrow, *The World Within the World*, Clarendon, Oxford (1988).

J.D. Barrow, *Theories of Everything*, Clarendon Press, Oxford (1991).

J.D. Barrow and F.J. Tipler, *The Anthropic Cosmological Principle*, OUP, Oxford (1986).

P. Davies, *God and the New Physics*, Pelican, Harmondsworth (1983).

P. Davies, *The Cosmic Blueprint*, Heineman, London (1987).

J. Glerk, *Chaos: Making a New Science*, Viking, New York (1987).

S.W. Hawking, *A Brief History of Time*, Bantam, London (1988).

R. Hooykass, *Religion and the Rise of Modern Science*, Scottish Academic Press, Edinburgh (1972).

J.T. Houghton, *Does God Play Dice?*—a look at the story of the Universe, IVP, Leicester (1988).

J. Maynard Smith, *The Problems of Biology*, OUP, Oxford (1986).

J.V. Narlikar, *The Primeval Universe*, OUP, Oxford (1988)

A.R. Peacocke, *Creation and the World of Science*, OUP, Oxford (1979).

A.R. Peacocke, *God and the New Biology*, J.M. Dent, London (1986).

R. Penrose, *The Emperor's New Mind*, OUP, New York (1989).

J.C. Polkinghorne, *One World*, SPCK, London (1986).

J.C. Polkinghorne, *Science and Creation*, SPCK, London (1988).

J.C. Polkinghorne, *Science and Providence*, SPCK, London (1989).

J.C. Polkinghorne, *Reason and Reality*, SPCK, London (1991).

C.A. Russell, *Cross-Currents: Interactions between science and faith*, IVP, Leicester (1985).

E. Squires, *Conscious Mind in the Physical World*, Adam Hilgar, Bristol (1990).

I. Stewart, *Does God Play Dice?*, Penguin, London (1990).

D. Wilkinson, *God, the Big Bang and Stephen Hawking*, Monarch, Tunbridge Wells (1993).